Sport Tourism Development

Sport Tourism Development

Maria Rellie B. Kalacas

SOCIETY Publishing

www.societypublishing.com

Sport Tourism Development

Maria Rellie B. Kalacas

Society Publishing
224 Shoreacres Road
Burlington, ON L7L 2H2
Canada
www.societypublishing.com
Email: orders@arclereducation.com

ISBN: 978-1-77469-142-7 (Hardcover)

ABOUT THE AUTHOR

Maria Rellie B. Kalacas is a licensed professional teacher and a mother of three respectful, endearing and God fearing children. She is also a dynamic researcher who has built her platform of healthy life balance and wellbeing on the back of thirteen year teaching career in Laguna State Polytechnic University Santa Cruz, Campus, Santa Cruz, Laguna, Philippines. She is holding the position of Assistant Professor III in the College of Industrial Technology department of Food Technology. She finished her Doctor of Philosophy major in School Leadership and Management at St. Dominic Savio College, Caloocan City, Philippines. She also completed her Master of Arts major in Technology and Home Economics from Laguna State Polytechnic University, Philippines. She also took up Master of Arts in Industrial Education major in Hotel and Restaurant Management at Eulogio "Amang" Rodriguez Institute of Science and Technology, Santa Mesa, Manila, Philippines. A member and awardee of Asian Community Social Response Humanitarian Ministry (ACSR) May 2018.With three Patented Utility Model research works and researches presented in the country.

TABLE OF CONTENTS

LIST OF FIGURES

LIST OF TABLES

LIST OF ABBREVIATIONS

ABIs	Area-Based Initiatives
ABS	Australian Bureau of Statistics
B2B	Business-to-Business
B2C	Business-to-Consumer
CCC	Canadian Climate Center
CPL	Cyberathlete Professional League
DMOs	Destination Management Organizations
EC	European Community
ECHAM	European Community and Hamburg Atmospheric Model
ESWC	Electronic Sports World Cup
ICT	Information and Communication Technologies
IOC	International Olympic Committee
IR	Interfirm Relations
KERS	Kinetic Energy Recovery System
LEDCs	Less Economically Developed Countries
MICES	Meetings, Incentives, Conventions, Exhibits, Shows, and Fairs
MPAs	Marine Protected Areas
NDC	New Deal for Communities
NGOs	Nongovernmental Organizations
NT	Networking Theory
NWDA	Northwest Development Agency
ROI	Return on Investment
SBU	Strategic Business Unit
SME	Small and Medium-Sized Enterprise
SRB	Single Regeneration Budget
TSA	Tourism Satellite Account
USPs	Unique Selling Points

WCG	World Cyber Games
WTO	World Tourism Organization

PREFACE

Tourism is one of the world's biggest and most lucrative sectors. It has a major societal impact, such as job creation, economic growth, and infrastructure development. Tourism and sports are inextricably linked and mutually beneficial. Sport-based tourism has the potential to develop into a tourism product with high economic returns for the country. The presence of large crowds and the participation of domestic and international athletes in sporting events have boosted the national economy. Tourism earnings have steadily increased year after year. This suggests that the tourism industry will continue to expand and potentially contribute significantly to the economy. Tourism and sports are the world's largest and fastest developing industries, both socially and economically.

Tourism generates a significant number of jobs in a variety of industries. These jobs may be in the tourism sector, but they may also be in the agricultural sector, the communication sector, the health sector, or the educational sector.

This book addresses in 8 chapters the central themes of sport tourism development. Chapter 1 is about the fundamentals of sport tourism that cover the basics of tourism and sports tourism. The 2nd chapter aims to explore the tourism industry in light of the most visited destinations, as well as the social, cultural, and geographical features that draw people to those destinations. The Olympic Games, the world's largest sporting event, provide significant brand value to the cities in which they are held. Tourism-related sporting activities are an appealing component for destinations, and when well-planned, they offer strategic benefits in destination management. Therefore, the 3rd chapter presents coverage on destination marketing and sports events management.

The 4th chapter aims to examine the scientific literature on tourism sociology as a subject to study the economic, social, and environmental impacts of tourism on societies and residents, as well as how residents perceived the benefits and costs of tourism developments in the local community.

Rural life can be combined with the manufacturing and service sectors to reduce unemployment. There must be certain prerequisites for tourism activities to occur in any area. There is a need for infrastructure and superstructure investments, which are essential for rural tourist attractions.

In the 5th chapter, the relationship of sport tourism activities in rural development is used to examine rural development and tourism styles. In rural areas, the positive and negative aspects of tourism activities are investigated. Chapter 6 is sports tourism and urban regeneration that has been a major theme for urban policy and planning in much of the developed world over the last three decades.

Chapter 7 aims to address the interdependence of sports tourism and sustainability from various disciplinary perspectives. Tourism is always changing, but there is a significant shift in demand and tourist expectations right now. Sports tourism brings together the best and worst of both the sports and tourism industries. Therefore, the global developments in sport and sports tourism will be shown first, based on the premise that social development is always a question of creative power and the influence of stakeholders. It is clear that the western, and increasingly commercialized, sports model is being exported globally. In the second step, current trends in sport tourism will be presented. Chapter 8 sheds light on current and future trends in sport tourism. Overall, highlighting the fields of sports events management, the sociology of sport, sports marketing, economic, urban, and sports geography, and tourism studies; this book will be a valuable resource for students, educators, and practitioners as well.

Chapter 1

Fundamentals of Sport Tourism

CONTENTS

1.1. INTRODUCTION

Tourism is a smokeless industry because there is no manufacturing process; income is produced through a variety of services provided to tourists. Tourism is a set of activities, services, and industries that provide a travel experience to individuals or groups of tourists who are away from home. It includes transportation, lodging, food, and beverages, shopping, entertainment, activities, and other hospitality services. Tourism is an industry that includes companies and organizations that provide tourists with services and features of tourist attractions. In this chapter, one tourism sector has been highlighted, namely tourism involving sporting activities. Tourism and sports are the world's largest and fastest developing industries, both socially and economically. Sports tourism encompasses all forms of active and passive participation in sporting activities, whether done casually or in an organized manner for noncommercial or business / commercial reasons that necessitate travel away from one's home and workplace. Major sporting events, such as the Olympic Games, football, and rugby championships, have become powerful tourism attractions in their own right, contributing significantly to the host destination's tourism image. Sports tourism refers to attractions that offer physical activities that entice visitors to spend a day in a specific tourism destination.

1.2. TOURISM BASICS

Tourism is one of the world's fastest-growing industries, contributing significantly to foreign exchange and job creation in many countries. It is one of the most extraordinary economic and social phenomena. The term "tour" comes from the Latin word tornus, which means "a tool for making a circle." Tourism is defined as the movement of people from their usual place of residence to another location for a minimum of 24 hours and a maximum of six months for the sole purpose of leisure and pleasure. As such, tourism is a product of modern social arrangements, having begun in Western Europe in the 17th century, though it has Classical antiquity antecedents.

According to the United Nations World Tourism Organization (UNWTO), tourism entails the movement of people to countries or places outside their usual environment for personal or business/professional purposes. These people are called visitors.

Tourism differs from exploration in that tourists adhere to a "beaten path," benefit from working systems of provision, and, as suits pleasure-seekers, are generally shielded from difficulty, danger, and embarrassment.

Tourism, on the other hand, intersects with other activities, interests, and processes, such as pilgrimage. As a result, shared categories such as "business tourism," "sports tourism," and "medical tourism" have emerged (international travel undertaken for the purpose of receiving medical care).

1.2.1. History of Travel and Tourism

Travel is as old as humanity on the planet. At the beginning of his existence, man roamed the earth's surface in search of food, shelter, security, and a better habitat. However, such movements were transformed into wanderlust over time.

Changes in climate, dwindling food and shelter conditions, and hostile invaders forced people to flee their homes and seek refuge elsewhere around 5000 years ago, similar to how the Aryans fled their homes in Central Asia due to climate change. This could lead to the growth of commerce, trade, and industry.

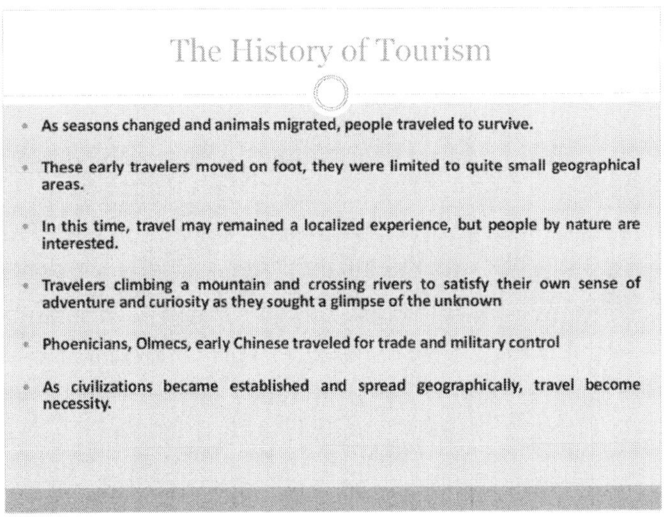

Figure 1.1. History of tourism.

Source: https://th.bing.com/th/id/Rf01856a3a9c4d653e0c6ec75518f1b9e?rik=D5qZXVoECrF40g&riu=http%3a%2f%2fimage.slidesharecdn.com%2fthehistoryoftourism-140415034811-phpapp01%2f95%2fthe-history-of-tourism-2-638.jpg%3fcb%3d1397533740&ehk=sCtrRGG9jgd6%2bkQM6NhJ%2blQqQdyftkldAg1fJq2aBlQ%3d&risl=&pid=ImgRaw.

During Hindu and Chinese civilizations, a religious, educational, and cultural movement began. Christian missionaries, Buddhist monks, and others traveled far and wide to deliver religious messages, returning with fantastic images and views about alien people (Figure 1.1).

For centuries, people's mobility increased due to the efficiency of transportation and the assistance and safety with which they could travel. By the end of the 15th century, Italy had established itself as Europe's intellectual and cultural center. It represented both the intelligentsia and the aristocracy's classical heritage.

During the 16th century, travel came to be regarded as an essential component of every young Englishman's education. Travel, in its broadest sense, became a means of self-development and education. The educational journey was dubbed the 'Grand Tour.'

The industrial revolution altered the pattern and structure of British society significantly. Thus, the British economy was largely responsible for the birth of modern tourism. It also resulted in a large and prosperous middle class, and as transportation systems improved in the latter half of the 18th century and the first quarter of the 19th century, a growing amount of people began to travel for pleasure.

Originally, travel was motivated by the need to survive (food, shelter, and security), the eagerness to expand trade, and the desire to conquer. As transportation improved, the desire to transform the vast and virgin world into a close neighbor spawned a new industry, namely travel and tourism.

However, the advancement of rails, roads, steamships, automobiles, and airplanes aided in the spread of technology around the world. Previously, travel was only available to the wealthy, but the industrial revolution changed everything. Transportation and lodging became more affordable to middle and working-class citizens. With the advancement of jet travel, communication, modern technology, tourism, and travel became the world's largest and fastest-growing industries. Travel and tourism have recently emerged as a dominant economic force on the global stage, contributing for more than 12% of total world trade and growing at an annual rate of 8%.

1.2.2. Nature of Tourism

Tourism, as a socioeconomic phenomenon, involves various activities and experiences of tourists and visitors away from home environment, as well as the services provided by the travel and tourism industry and host destination. This activity experience and services can be viewed as a tourism product.

The tourism system can be divided into two parts: supply and demand. Tourism planning should strive for a balance of supply and demand. This necessitates not only an understanding of market characteristics and trends, but also of the planning process used to meet market demands.

The demand side is frequently identified as tourists from core generating markets; the supply side includes all facilities, programs, attractions, and land uses designed and managed for visitors. These supply-side factors may be managed by private enterprise, non-profit organizations, or the Government. New and innovative forms of collaboration are also emerging to ensure the long-term development and management of tourism resources.

The supply and demand sides are linked by flows of resources such as capital, labor, goods, and tourist expenditures into the destination, as well as flows of marketing, promotion, tourist artifacts, and experiences from the destination back into the tourist-generating region.

Furthermore, some tourist expenditures may leak back into visitor-generating areas as a result of foreign tourism investors' profits being repatriated and payment for enhanced goods and services provided to tourists at the destination. Transportation is essential for getting to and from the destination.

The major supply-side components for planning purposes are:

- Various modes of transportation and other tourism-related infrastructure;
- Tourist information;
- Marketing and promotion;
- The community of communities within the tourist's destination area;
- The political and institutional frameworks for facilitating tourism.

Because of the many factors that are linked to it and the existence of many sectors that contribute to its success, the tourism system is both dynamic and complex. These factors and sectors are related to the provision of a tourist experience as well as the generation of tourism revenue and markets.

The dynamic nature of the tourism system necessitates a regular scan of the external and internal environments of the destinations in order to make changes as needed to maintain a healthy and viable tourism industry.

As a result, it is now widely acknowledged that tourism development cannot take place in isolation from the environment and local communities, nor can it overlook the social and cultural implications of tourism.

1.2.3. Purpose of Tourism

Tourism is critical to the success of many economies worldwide. Tourism has a number of advantages for host cities. Tourism increases the economy's revenue, creates thousands of jobs, develops a country's infrastructure, and fosters a sense of cultural exchange between foreigners and citizens.

Tourism generates a significant number of jobs in a variety of industries. These jobs may be in the tourism sector, but they may also be in the agricultural sector, the communication sector, the health sector, or the educational sector. Many tourists travel to experience the culture, various traditions, and gastronomy of the host country. Local restaurants, shopping centers, and stores benefit greatly from this (Figure 1.2).

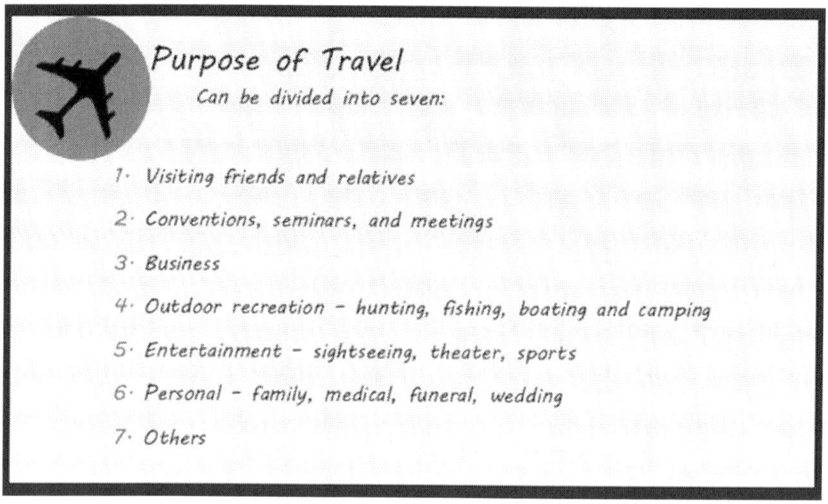

Figure 1.2. Different purposes of travel.

Source: https://image.slidesharecdn.com/meaningandimportanceoftourism-150419190735-conversion-gate02/95/meaning-and-importance-of-tourism-18-638.jpg?cb=1429488622.

Governments that rely heavily on tourism for a large portion of their revenue invest heavily in the country's infrastructure. They want more tourists to visit their country, which necessitates safe and advanced facilities. This results in new roads and highways, parks, better public spaces, new airports, and considerably better schools and hospitals. Infrastructure that is both safe and innovative allows for easy flow of goods and services. Furthermore, local residents benefit from opportunities for economic and educational growth.

Tourism fosters cultural exchange between tourists and locals. Foreigners are typically drawn to exhibitions, conferences, and events. Organizing authorities typically profit from registration fees, gift sales, exhibition space sales, and media copyright sales. Furthermore, foreign tourists contribute to the host country's diversity and cultural enrichment.

Tourism provides an excellent opportunity for foreigners to learn about a new culture, but it also provides numerous opportunities for local residents. It enables young entrepreneurs to launch new products and services that would not be viable on the basis of the local population alone. Furthermore, residents benefit from tourism that takes place in their own country.

1.2.3.1. What Is the Main Purpose of Tourism?

1. *Economic Sustainability:* It ensures the viability and competitiveness of tourism destinations and businesses. It promotes continuous improvement, which is advantageous in the long run.
2. *Local Enrichment:* The prosperity of tourist destinations is a huge part of tourism. The tourism industry has consistently maximized the host destination's economic growth.
3. *Employment Standards:* Tourism promoted a high level of salaries, service terms, and accessibility for all. It creates employment in the community with no discrimination based on gender, ethnicity, disability, or other factors.
4. *Local Management:* Include local groups and encourage them to plan and make decisions. This was developed with the assistance of the neighborhood tourism management team.
5. *Public Welfare:* Maintain and improve the lifestyle of the local community. They are a component of social systems. But, without a doubt, this procedure should take place without any social embarrassment or exploitation.
6. *Natural Integration:* Preserve and enhance the nature of both urban and rural environments. It prevents both natural and visible environmental degradation.
7. *Biodiversity:* Another aim of tourism is to help with the protection of habitats and natural wildlife areas, as well as to minimize losses.

8. *Environmental Cleanliness:* In addition to the intent of tourism, all visitors must minimize air, water, soil, and waste pollution.
9. Tourism is an important component of national integration.
10. Tourism often encourages visitors to learn about their customs, history, culture, and religious beliefs.
11. The most important aspect of tourism is economic development or the destination's industry. It encourages locals to make a variety of handicraft products, as well as local foods, souvenirs, clothing, and other items for sale.
12. Tourism is one such industry that provides a consistent economic boost to the nation.

1.2.4. Importance of Tourism

Tourism and hospitality, which are deeply intertwined, are two of the world's most profitable industries. They are also among the top employers. As travel has become more common, there has been an upmarket trend in tourism over the last few decades. People travel for a variety of reasons, including business, pleasure, adventure, and even medical treatment.

With many business-related activities linked with tourism, the industry has enormous potential for both job creation and foreign exchange earnings. Many countries around the world, including Mauritius, Malaysia, Singapore, Fiji, and the Caribbean, rely heavily on tourism to fuel their economies. Tourism can help a country's economic growth in the following ways:

1. ***Creation of Employment:*** It generates a large number of jobs for both direct service providers (like hotels, restaurants, travel agencies, tour operators, guide, and tour escorts, and so on) and indirect service providers (such as suppliers to the hotels and restaurants, supplementary accommodation, etc.).
2. ***Infrastructure Improvement:*** Tourism encourages the development of infrastructure. Any location would require all of the necessary infrastructure, such as good connectivity through rail, road, and air transport, sufficient accommodation, restaurants, a very well telecommunication network, and medical facilities, among other things, in order to become an important commercial or pleasure destination.
3. ***Foreign Exchange:*** People who travel to other countries spend a lot of money on things like lodging, transportation, sightseeing,

shopping, and so on. As a result, an inbound tourist is a valuable source of foreign exchange for any country.

1.2.5. Industries Related to Tourism

Tourism has grown in popularity around the world over the years. Tourists require and demand specific facilities and services based on the nature and purpose of their trip. This has resulted in a plethora of commercial activities that have grown to industry proportions. As a result, travel and tourism now encompass a wide range of related industries:

1. ***Hotels:*** These are commercial establishments that offer lodging, meals, and other guest services. The hotel industry is very important in the travel and tourism industry because all tourists need a place to stay at their destinations, as well as many other services and facilities to accommodate their specific needs and tastes.
2. ***Restaurants:*** These are retail establishments that serve customers prepared food and beverages. Restaurants and other food and beverage outlets are very important in the travel and tourism industry because tourists like to try the local cuisines of the places they visit.
3. ***Retail and Shopping:*** The retail industry is vital because tourists shop for daily necessities as well as mementos and souvenirs. In recent years, some cities around the world have been promoted as shopping destinations in order to attract people who enjoy shopping by offering a variety of products such as clothing, electronics, jewelry, and antiques. Fashion capitals of the world include New York, Paris, London, and Milan, Italy.

   The movement of people and goods from one location to another is referred to as transportation. A well-developed transportation industry, as well as infrastructure, is critical to the success of any travel and tourism business.
4. ***Travel Agencies:*** It is a retail establishment that sells travel-related services and products, especially package tours, to customers on behalf of suppliers, including airlines, car rentals, cruise liners, hotels, railways, and sightseeing.

   Travel agencies play an important role because they plan their clients' itineraries and make the required accommodations for their travel, stay, and sightseeing, in addition to facilitating their passport, visa, and so on.

5. ***Tour Operators:*** To create a holiday, it typically combines tour and travel components. Tour operators are vital to the travel and tourism industry.
6. ***Tourist Destinations:*** A tourist attraction is a location of interest to tourists, typically because of its inherent or demonstrated cultural value, cultural importance, natural or built beauty, or amusement opportunities. These are the fundamental principles of the tourism industry.
7. ***Cultural Industries:*** Cultural or creative industries are in charge of the creation, production, and distribution of cultural goods and services, which are typically protected by intellectual property rights. Because tourists enjoy visiting culturally significant sites and immersing themselves in the local culture, the cultural industry is critical to travel and tourism.
8. ***Sport, Recreation, and Leisure:*** Leisure, also known as free time, is a period of time spent away from work and essential domestic activities. Spending time in a way designed for therapeutic refreshment of the body or mind is referred to as recreation or fun. While leisure is more akin to entertainment or rest, recreation necessitates active involvement in a refreshing and diverting way.

The need for recreation has grown as people in the world's wealthier regions have become more sedentary. These play an important role in the travel and tourism industry.

1.3. TOURISM PRODUCTS

A tourism/tourist product is referred to as the sum of the physical and psychological satisfaction it provides to tourists while they are en-route to their destinations.

Because the travel and tourism industry is comprised of numerous sectors that promote travel-related services. These industries are known as travel vendors, and their goods and services are known as 'travel products.' Physical plant, services, hospitality, freedom of choice, and a sense of involvement are the five main components of a tourism product.

Thus, whatever man-made and natural resources, services brought about by tourist consumption are referred to as tourism products.

1.3.1. Characteristics of Tourism Products

You have probably figured out what a tourism product is by now. Let us now examine some of its characteristics (Figure 1.3):

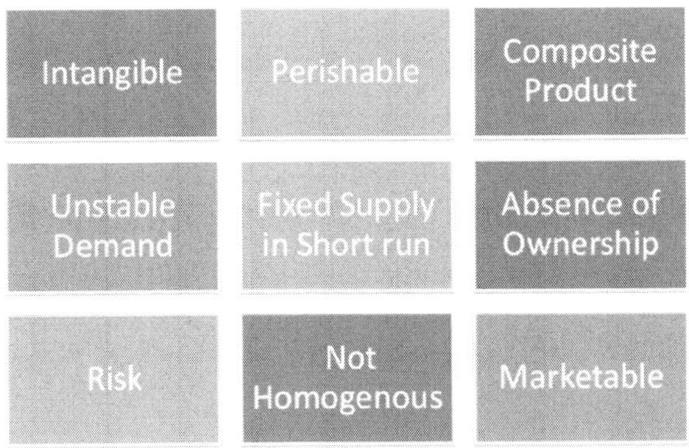

Figure 1.3. Characteristics of tourism products.

1. ***Intangible:*** Tourism is an intangible product, which means it cannot be touched or seen, and there is no transfer of ownership. However, the facilities are available for a limited time and for a limited use. For example, a hotel room is available for a set period of time.

2. ***Psychological:*** The primary reason for purchasing a tourism product is to satisfy a psychological need after using the product, which is satisfied by gaining experience while interacting with a new environment. In addition, personal experiences influence others to buy a product.

3. ***Highly Perishable:*** Because tourism products are highly perishable, they cannot be stored for an extended period of time. While a tourist is present, production, and consumption occur. If the product remains unused, i.e., if tourists do not purchase it, the chances are lost.

A tourist product cannot be stored by a travel agent or tourism operator who sells it. Production can take place only if the customer is present. Furthermore, once consumption begins, it cannot be stopped, interrupted, or altered. If the product remains unused, the opportunity is lost; for example, if tourists do not visit a specific location, the opportunity at that time is lost. Because of the tourism industry, hotels, and transportation companies offer significant discounts during the offseason.

4. ***Composite Product:*** A tourist product is a collection of various products. It is not a single entity in and of itself. Various service providers, such as transportation, contribute to the experience of a visit to a specific location. Unlike a manufactured product, the tourist product cannot be provided by a single enterprise.

   The tourist product encompasses the entire experience of visiting a specific location. In addition, many service providers contribute to the tourism experience. For example, an airline provides seats, a hotel provides rooms and restaurants, and travel agents make reservations for lodging and sightseeing, and so on.

5. ***Unstable Demand:*** The tourism market is influenced by seasonal, economic, and political factors, among others. There are times of the year when demand is higher than others. There is a greater strain on services such as hotel bookings, employment, and the transportation system at this time.

1.4. TECHNOLOGY AND THE DEMOCRATIZATION OF INTERNATIONAL TOURISM

Transport innovation was a critical enabler of tourism's spread and democratization, as well as its eventual globalization. Beginning in the mid-19th century, the steamship and railway provided greater comfort and speed, as well as cheaper travel, due in part to fewer overnight and intermediate stops. Above all, these advances allowed consistent time-tabling, which was critical for those who relied on the calendar, if not the clock. The accessibility gaps to these transportation networks were slowly closing in the late 19th century, as the steam empire expanded globally. Railways promoted both domestic and foreign tourism, including short trips to the coast, city, and countryside that lasted less than a day but clearly qualified as "tourism." Rail travel also increased the accessibility of grand tour destinations, strengthening existing tourism flows while leading to tensions and conflicts between classes and cultures among tourists. Steam navigation and railways were opening tourist destinations from Lapland to New Zealand by the late 19th century, with the latter opening the first dedicated national tourist office in 1901. Governments became involved in tourism after WWII as an invisible import and a weapon of diplomacy, but prior to this period, global travel agencies took the lead in easing the complexities of tourist journeys.

The most well-known of these organizations was Britain's Thomas Cook and Son, whose activities spread across the globe from Europe and the Middle East in the late 19th century. Other firms' roles have been less evident to 21st-century observers, owing in part to the fact that these agencies did not maintain their records, but they were equally significant. From the late 19th century onward, shipping lines encouraged foreign tourism. Before World War I, pleasure cruises were becoming a distinct tourist activity from the Norwegian fjords to the Caribbean, and transatlantic companies competed for middle-class tourism during the 1920s and 1930s. Between the World Wars, wealthy Americans traveled by air and sea to a number of Caribbean and Latin American destinations.

Tourism grew much larger on a global scale in the latter half of the 20th century, as air travel was gradually deregulated and decoupled from "flag carriers" (national airlines). The airborne package tour to sunny coastal destinations became the foundation of a massive annual migration from northern Europe to the Mediterranean, before spreading to an increasing number of long-haul destinations, including Asian markets in the Pacific, and finally bringing post-communist Russians and eastern Europeans to the Mediterranean. Similarly, traffic flows from the United States to Mexico and the Caribbean increased. In each case, these advancements were based on older rail, road, and sea travel trends. During the 1930s and postwar years, the first package tours to the Mediterranean were by motor coach (bus). It was not until the late 1970s that Mediterranean sun and sea vacations became common among working-class families in northern Europe; the term "mass tourism," which is sometimes used to describe this phenomenon, is inaccurate. Since visitors had options, such vacations were enjoyed in a number of ways, and the destination resorts varied greatly in history, culture, architecture, and visitor mix. The rise of budget airlines, especially EasyJet and Ryanair in Europe, opened up a new mix of destinations beginning in the 1990s, allowing for more flexible international travel. Some of these were former Soviet-bloc destinations like Prague and Riga, which drew weekend and short-break European tourists who developed their own itineraries through negotiations with local service providers mediated by airline special deals. Globalization in international tourism has not been a one-way street; it has necessitated negotiation between hosts and visitors.

1.5. DAY-TRIPPERS AND DOMESTIC TOURISM

Although domestic tourism may appear to be less glamorous and dramatic than international traffic flows, it has become more relevant to more people over time. Since the 1920s, the emergence of Florida as a tourist destination for Americans has been marked by "snowbirds" from the northern and Midwestern states traveling a greater distance throughout the vast expanse of the United States than other European tourists travel internationally. Domestic demand and local journeys drove key phases in the innovative growth of tourism as a commercial phenomenon in the United Kingdom. European wars in the late 18th and early 19th centuries sparked the "discovery of Britain," as well as the emergence of the Lake District and Scottish Highlands as destinations for both the upper and aspiring classes. The railways, especially in the last quarter of the 19th century, aided in opening up the seaside to working-class day-trippers and holidaymakers. By 1914, Blackpool in Lancashire, the world's first working-class seaside destination, had about 4 million summer tourists. Coney Island in Brooklyn, New York, had more tourists by this time, but the majority were day-trippers who came from and returned to other locations in the New York City area by train on the same day. Domestic tourism is less evident statistically and is typically served by regional, local, and small family-run businesses. The World Tourism Organization (WTO), which attempts to count tourists globally, is more connected with the international scene, but domestic tourism remains much more significant in numerical terms than international tourism around the world, perhaps particularly in Asia.

1.6. SPORT TOURISM: AN OVERVIEW

Individuals traveling to a destination to compete in or watch sports such as world games, international test series, stadium tours, mass participation activities, player testimonials, community tours, sporting events, and exhibitions are examples of sports tourism. Sports tourism is an important component of global tourism. Travelers are interested in sports tourism and want to learn about different destinations that host globally acclaimed sporting events. The Olympics, F1 Grand Prix, FIFA World Cup, Tennis Grand Slams, PGA Championships, and Cricket World Cup are all outlets for promoting the destination. Individuals from all ages and backgrounds are interested in sports tourism. There has been a significant rise in the number of visitors traveling to watch international sporting events over the years. Sports tourism is classified into two types: spectator sports events, where people come to watch the game, and participatory sports events, where people come to engage in the activities (Sport Tourism, 2014). Sports tourism is described as tourism planned by individuals with the goal of watching or participating in various sporting events. It is a form of tourism that focuses on sports. Individuals all over the world are developing a strong interest and passion for various sports and want to pursue them as a profession. In most cases, they cultivate an interest in sports at a young age, begin practicing, and by the time they reach adolescence, they are competent enough in the sport. When people develop a strong interest in a specific sport and intend to pursue it as a career, it is important that they engage in professional training and participate in sporting events. Individuals' specific travel outside of their home region is referred to as sports tourism. Individuals' participation in the sporting event can be active or passive. In this situation, when people schedule a trip from one location to another, sport is the primary reason. Individuals schedule sports tourism in certain situations where they need to observe events or tournaments in which their relatives or friends are competing. Leisure and entertainment components can be included in this tourism in order to enhance the sense of sports tourism. Sports tourism is typically divided into two different categories. There are two types of sports tourism: active sports tourism and passive sports tourism. Traveling for the purpose of engaging in a sport, leisure, or recreational activity is referred to as active sports tourism. Passive sports tourism, on the other hand, is travel for the purpose of visiting a sport, leisure, or recreational activity (Figure 1.4).

Figure 1.4. Sports tourism is an increasingly growing segment of the global travel industry, with an annual revenue of $600 billion. It has been subdivided into notable products such as golf tourism, polo tourism, and adventure tourism through various activities.

Source: https://webunwto.s3.eu-west-1.amazonaws.com/2020-01/sport-congresse.jpg.

1.6.1. Classification of Sport Tourism

Sports tourism is classified into many categories. According to one theory, sports tourism can be divided into two types: hard sports tourism and soft sports tourism. According to the other theory, there are three kinds of sports tourism: sports event tourism, celebrity and nostalgia sports tourism, active sports tourism, inbound sports tourism, and outbound sports tourism.

1. ***Hard and Soft Sport Tourism:*** A hard concept of sports tourism is the number of people who attend professional sporting events. In general, these types of events are the stimuli that entice visitors to attend these events. Olympic Games, FIFA World Cup, Formula One Grand Prix, and regional competitions such as the NASCAR Sprint Cup Series are examples of hard sports tourism. Soft definition, on the other hand, is moderately the travel of tourists for the purpose of participating in sporting events that are organized primarily for recreation purposes or signing up for leisure interests. Hiking, skiing, and canoeing are examples of soft sports tourism.
2. ***Sports Event Tourism:*** It refers to individuals who schedule a trip to a city with the intention of attending sporting events. In most cases, a significant number of people cultivate an interest in sports. When there are sporting activities or matches, people normally get excited and enjoy watching them. In certain cases, they may also reduce their working hours to attend a sporting event. During the Olympics, this is a prime example. Each Olympic host city receives a massive influx of tourists. Simply put, when a sporting event is scheduled, people plan trips to the city or even another country.
3. ***Nostalgia Sport Tourism:*** It entails visiting well-known and prominent sports-related destinations. Individuals are typically impressed by tournaments and the venues where sporting events are held. When they raise awareness of the location of the sporting event, they usually develop motivation to visit the location and plan their visit. Visits to well-known sporting halls such as the Women's College Basketball Hall of Fame in Knoxville, Tennessee, and the Baseball Hall of Fame in Cooperstown, New York are suitable examples of these sports-related formations. This category also includes sports museums, such as the NASCAR museum in Charlotte, North Carolina, and

well-known sports venues, such as Lambeau Field in Green Bay, Wisconsin, and so on. Individuals who travel primarily for the purpose of participating in sports events are referred to as active sports tourists. These participatory events can take many different forms in a wide range of sports. Individuals have developed a keen interest and enthusiasm for sporting events in the modern era, as well as in the ancient past. Participation in active sports tourism requires extensive preparation, which is one of the most important aspects. Individuals in most cases need to acquire knowledge and information, as well as develop their skills and capabilities, through the acquiring of training from a professional trainer. Golf, kayaking, tennis, fishing, cricket, swimming, and surfing are just a few of the sports that people travel to participate in.

4. ***Inbound Sport Tourism:*** Adventure sports and games are at the heart of sports tourism. There are undeniably numerous locations throughout the country that contribute to sports tourism. During a vacation, one can participate in a variety of sports activities. Mountaineering, rock climbing, scuba diving, water rafting, kayaking, canoeing, sailing, surfing, and water scooting are examples of adventure sports. Ballooning, paragliding, and hand gliding are examples of aero sports. These sports provide opportunities for sports lovers from all over the world.
5. ***Outbound Sport Tourism:*** In most situations, people schedule a vacation with their friends and family in order to attend a sporting event. There has been an increase in the number of sports fans who spend substantial sums of money on trips to watch sporting events in the modern era. Outbound sports tourism refers to tourism that includes traveling to locations outside of the country. In today's world, there are a large number of visitors who fly abroad to see a variety of sporting activities. Cricket, soccer, tennis, badminton, wrestling, and another sport fall under this category. Although the majority of sports tourism inside the country is outbound, there are some minor inroads being made on the inbound and domestic fronts. According to industry analysts, the segment is expected to expand at a pace of 10 to 20% in the future.

1.6.2. Characteristics of Sport Tourists

The characteristics of sports tourists have been stated as follows:

1. ***Interest and Enthusiasm:*** Sports tourists typically cultivate an interest in different things that may or may not be taken into account by other people. They are usually health-conscious and participate in a variety of physical activities such as hiking, yoga, and so on. Individuals who participate in sports are typically both males and females, they are affluent, and they are engaged not only in sports but also in other fields, such as the execution of household chores and the performance of job duties. These people may or may not have a college education. Athletes have been studied extensively, and it has been discovered that while people are sports fans, they would normally get up early in the morning in order to participate in practice sessions. Individuals must properly prepare themselves, particularly when participating in a sporting event (Ross, 2001).
2. ***Traveling by Themselves:*** Traveling in certain cases is not safe, especially for the women. There has been a rise in the number of women of different ages who have been exposed to illegal and abusive acts, and there has been a prevalence of crime and violence. One of the significant characteristics is that when individuals are invited to compete in a sporting event, they must travel from one location to another and bear all expenses. Individuals may travel with friends or family members in some cases but must travel on their own in others. As a result, these people are generally eager and interested in planning their travel from one location to another. They are generally aware of the aspects that are used to resolve any kind of insecurity and apprehension that might occur during their expedition.
3. ***Health and Happiness:*** Sports visitors, for the most part, pay close attention to their fitness and well-being. They are professional and aware of all aspects needed to properly care for one's well-being. Maintaining good health and being free of different forms of diseases and health issues is critical for these people. Getting involved in physical activity, eating a balanced and nutritious diet, keeping stress and tension-free, and developing positive thinking are some of the different things that they usually consider in order to maintain good health. It is also important for individuals to

get 6 to 8 hours of sleep a day. It is critical that they implement a variety of steps in order to gain the energy and dynamism needed to engage in sports tourism.

4. ***Discipline:*** It is important for individuals to be disciplined in order to achieve success in task completion. In most cases, when sports enthusiasts participate in practice sessions, they must be disciplined and effective. Training and preparation are usually carried out under the guidance of a trained and knowledgeable teacher. An ideal candidate will be someone who is well-versed in the sport and has prior experience. Typically, coaches who are interested in delivering instruction to individuals have a strict mindset. It means that they allow them to participate in practice sessions early in the morning that they force them to participate in vigorous practice exercises that they force them to eat a balanced and nutritious diet, which they force them to give up some food products, and so on. These factors help us understand that successful sports visitors must adhere to a strict schedule and give up some things.

5. ***Effective Communication:*** Sports fans may or may not be prepared. They may have a college degree, a high school or middle school diploma, or only basic reading skills. Individuals acquire strong communication skills by engaging in athletic activities and traveling to various locations. Since they must communicate with a large number of people, it is important for them to be respectful and ethical in their contact with others. There have been reports of athletes who are introverted and unable to engage with others. In certain cases, people have negative beliefs that if they engage with others, they will be affected or suffer negative consequences. As a result, these individuals generally feel more at ease interacting with others they meet. Participating in sports and attracting attention does make a huge contribution to instilling traits of righteousness, ethics, and morality in them.

6. ***Diligence and Resourcefulness:*** In some situations, people cultivate an interest and passion for engaging in sporting activities as early as adolescence. Individuals typically cultivate an interest in studies, career prospects, and sports events. Individuals are normally exposed to different sports from an early age. In other words, as they participate in pre-school, they are educated in a variety of sports such as tennis, badminton, running, water

sports, and so on. When they enter regular schools, they begin to participate in competitions and tournaments, particularly if they develop an interest and skills in a specific sport. Individuals who participate in sports tourism will cultivate qualities such as diligence and resourcefulness. These characteristics are most noticeable when performing other activities and roles, such as learning, working, and so on. Individuals are expected to carry out other activities and roles throughout their everyday lives, which necessitates responsible conduct. This includes household management, taking care of the needs and requirements of other family members, and so on. Participating in a variety of activities, or even just one, helps to instill healthy behavior in individuals. Individuals are able to understand their obligations to their family members and contribute significantly to their well-being. In certain cases, these people engage in sports competitions not only to improve the prestige and popularity of their communities, but also of the society as a whole. When individuals travel to other countries to compete in sporting competitions, they are ultimately responsible for representing the entire country.

7. ***Adaptable Nature:*** One of the most significant attributes of sports tourists is their adaptability. They are easily adaptable to changing environments and situations or circumstances. These conditions and circumstances may be linked to sporting events or to other issues. These abilities form a sport to measure for each, for students on educational tours that can be structured; for athletes who want to cope with a task in which speed and endurance are combined with the power of choice and the speed at having to make decisions, particularly for the elderly, who can be fitted with a motor task and knowledgeable who is, at the same time, even considered to be a healthy competition (Giorgio and Spinelli, n.d.).

8. ***Development of Intellectual Abilities:*** When these people actively participate in sporting activities or schedule trips solely for the purpose of observance, they are working on the development of their intellectual abilities. These skills are typically established as people use the internet to search for different forms of sports and knowledge about them. They typically become engrossed in reading various sports-related texts and materials, and they may even grow an interest in training. There are numerous

training institutes located across the country that help to prepare individuals. Numerous studies on sports events have been conducted, with the majority of them focusing primarily on the measurement of economic impacts caused by so-called mega-events (Giorgio and Spinelli, n.d.).

9. ***Marketable Development:*** Sports tourism plays an important role in the marketable and financial development of individuals. Active sports visitors work hard, and in some cases, they must put in long hours of practice in order to foster competence and skills. Participation in sports activities, along with the acquisition of skills and expertise, makes an important contribution to an individual's marketable development. Sports tourism is a new product in the global tourism industry that is dependent on sports capital. Sports tools are used in different types of sports activities for the purpose of organizing, structuring, designing, and combining as relevant items that can stimulate individuals' consumption desire and demand. Meanwhile, it is a modern form that will enable individuals to engage in sports and gain natural experiences. There has been a great deal of academic interest in sports tourism, with a focus on marketable development (Wickramaratne and Kumari, 2016).

1.6.3. Benefits of Sport Tourism

Sports tourism is a significant component of global tourism and has evolved into a highly competitive niche market. As people's interest and participation in sports grow in the leisure industry, so does their desire to travel to compete or spectate. Sports tourism entails more than just staging spectacular activities.

Individuals of all ages, from all socioeconomic backgrounds and socioeconomic divisions, have benefited from sports tourism. The following are the primary advantages of this tourism:

1. ***Economic Growth:*** Sports are an investment in the tourism sector, which results in successful economic development. They stimulate economic growth by providing various types of lodging, such as hotels, restaurants, and retail establishments. In cities or regions where sports activities are organized, attempts are made to lead to changes in different areas that may make an effective contribution in promoting the visit of individuals (Figure 1.5).

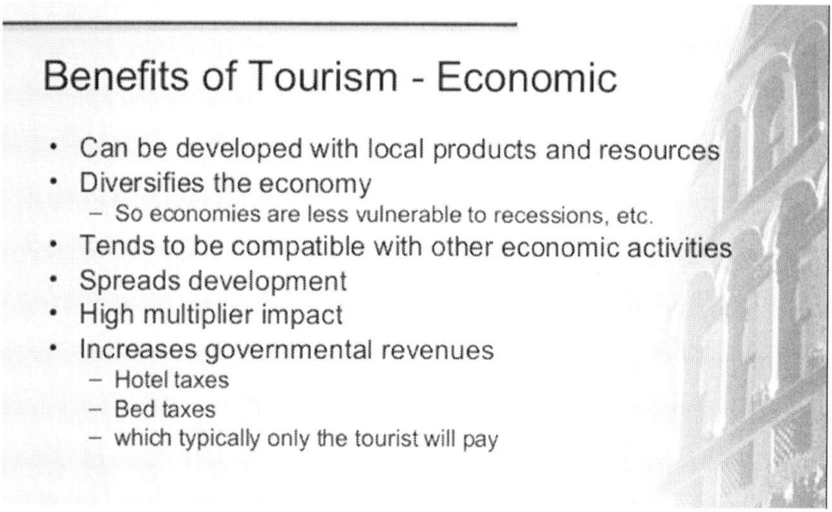

Figure 1.5. Economic impacts of tourism.

Source: https://image.slidesharecdn.com/lecture2impactsoftourism-globaldist-130516181251-phpapp01/95/lecture-2-impacts-of-tourism-global-dist-31-638.jpg?cb=1368728050.

2. ***Experience:*** Sports tourism creates experiences for individuals while still developing a positive picture for the community. Individuals who have a strong interest in a specific form of sport and a strong desire to learn it normally plan sports tourism. Sports tourism will allow them to watch or participate in the sport, thereby enhancing their experience. They are able to recognize faults and contradictions and make changes as a result. Individuals' experiences are enhanced when they participate in sporting activities on a regular basis. They, on the other hand, will develop proficiency and competence over time.
3. ***Product Development:*** Sports tourism creates a new product as well as a new tourism destination. When the word "product" is used, it refers to the sport to which one devotes sufficient attention. Individuals who plan sports tourism solely for the purpose of attending the event then obtain specialized training to improve their skills and abilities, which is referred to as product development. Furthermore, when product development occurs, one is able to make the best use of the community's abilities, especially expertise, skills, encouragement, and enthusiasm.

4. ***Community Relations:*** Sports tourism fosters community relationships while also reinforcing corporate sponsorship. Individuals who are committed to a specific sport, are athletes by profession, who engage in different competitions and activities build group relationships. As a result, the individuals improve their contact ties and build relationships and terms with the other people. Such individuals can include promoters, other athletes, and those participating in sporting events. There have been examples of people who have been involved in sporting competitions since they were adolescents. In certain cases, they begin learning as early as adolescence, and since they practice the sport from the beginning, they are able to become proficient as they reach adulthood. During this point, they participate in a variety of national and international events, allowing them to improve their status and credibility while also developing community relationships. Sports tourism is recognized as a significant source of entertainment for young people. Individuals must focus when participating and exhibit qualities such as patience, resourcefulness, and conscientiousness. Furthermore, they gain the ability to improve their image while still providing entertainment. Individuals who are mainly watching sporting events, on the other hand, are entertained and amused. When matches between two players or two teams are organized, most people have a favorite team based on their ethnicity or other variables, and they are amused if their favorite team wins. Make a drawing High rate of return Visitors-The most profitable visitors are those who have already visited. Individuals have previously visited that location, either for the purpose of involvement or for observation. They are familiar with the location, and each time they scheduled a visit, their primary goal was to either observe or participate in the sporting event. As a result, one of the advantages of organizing sporting activities is that tourists schedule their trips to the city again. As a result, the city attracts high-yield tourists.
5. ***Provides a Strong Image of the Destination:*** Individuals are able to create positive perceptions of tourist destinations, especially those to which they plan sports tourism on a regular basis. On the other hand, when people are required to visit a location on a regular basis, especially for the purpose of engaging in sports, they are able to create a positive image for the location. They do

not pay attention to these areas if the region is heavily populated or if the infrastructure is not well established, largely because of their purpose, and they form a positive picture for the destination.

6. ***Infrastructure:*** The development of modern infrastructure is a critical component in advancing sports tourism. Infrastructure is a general concept that encompasses public utilities, communications, public transportation, power, and water sources, telecommunications, radio, and television broadcasting, roads, and rail networks, and overall organization. When a sporting event is held in a community, it is critical to ensure that all areas are well-developed. Individuals will be able to benefit from sports tourism when infrastructure facilities are well-developed and well-maintained.
7. ***Communication:*** It is known as the most important component of a sporting event organization. For example, during a cricket match, the function of the commentators is to relay important information to the audience. In this way, one can improve their understanding of the scores, overs, number of balls available, and so on. A large number of people watch or listen to these activities on television or on the radio. Nowadays, with the advancement of technology, people watch these matches and events on their cell phones, smartphones, and so on. The use of the media to effectively interact is regarded as an important feature of sports.
8. ***Growth of Sport Tourism:*** The growth of infrastructure, connectivity, and individual participation and excitement contribute to a rise in the rate of sports tourism. Individuals who are planning sports tourism to another city or country can also participate in other leisure and entertainment activities. As a result, they will schedule their sports tourism and leisure tourism at the same time. Furthermore, sports tourism can plan organizational marketing and the community's bidding power.

1.6.4. Factors that Influence Sport Tourism

The factors that affect sports tourism can be divided into three categories: economic, technological, and socio-cultural:

1. ***Economic Factors:*** A sufficient amount of money is needed for individuals to engage in any form of sports tourism. Families' wealth in the present has reached new heights, owing primarily

to significant shifts in the types of job prospects available to individuals. The amount of leisure time available for sports tourism activities is increasing for individuals. In general, the number of hours worked per week has decreased over the last century, especially as the number of holidays has increased and technological advancements have resulted in more time for many people. Finally, for most Western societies, economic developments have increased the amount of time and resources available to people, motivating them to participate in sports tourism activities.

2. ***Technological Factors:*** Recent technological advancements are also significant in enabling the expansion of sports tourism to a broader cross-section of the population. New transportation approaches are the best example of new technologies. People have been able to fly to different locations to engage in sports tourism events since the advent of cars and air travel in the previous century. Furthermore, with the development of modern transportation methods, people started to seek out lodgings away from their homes. This resulted in a huge increase in the number of hotels, inns, and motels. Aside from the role that transportation and lodging played in growing the popularity of sports tourism, new developments in media technology also aided in growth. The newspaper, radio, and television all played an important role in the growth of sports tourism. Finally, advancements in sports equipment manufacturing have aided in the growth of sports.
3. ***Socio-Cultural Factors:*** Sports tourism will strengthen national heritage, individuality, uniqueness, and community spirit by bringing together locals to promote their culture. Sports tourism can provide a vehicle for tourists to learn about people from other countries and their cultures. Sports tourism will help to restore and preserve cultural traditions.

1.6.5. The Development of Sport Tourism Industry

The sport evolved during the late 19th and early 20th centuries (1880–1910). Sports were changed through such breakthroughs as the codification of rules, bureaucratization (the development and regulation of competitions), and diversification of player roles, measurement of performance, and the maintenance of records of accomplishment (Guttmann, 1974, in Higham,

2005). Local, regional, and national leagues were used to compete in sports.

The second wave of dynamic change in the history of sport occurred in the 1970s and 1980s. (Halberstam, 1999; in Higham, 2005) investigates the rise of cable television and the coverage of live sporting events, which began with ESPN in 1978. The growth of commercial and media interest in sport, as well as the rise of new types of sports celebrity, resulted in sport tourism becoming a dominant feature. As a result, there is a rising demand for sports-themed vacations.

Sport, sports teams, sporting events, and sporting facilities have all been turned into the tourism industry. Sport is a vital cultural expression in a tourist destination. Sport has become such an influential weapon in destination marketing due to its popularity in the national and international media (British Tourist Authority, 2000; in Higham, 2005).

1.6.5.1. Market Analysis

We can see an increase in tourist arrivals all over the world over the years. It is regarded as a significant phenomenon that will both engage and excite people. The focus, for example, here will be on sports tourism in Australia. Sports tourism makes a significant contribution to the Australian economy.

Australia is known around the world as a sports-crazed country. According to the Australian Government (2008), the total expenditure on recreation, cultural, and sports services was $2,953 million. The Australian Government invested $3.8 billion in arts and recreation services in 2008–2009. This demonstrates that they are constantly improving recreation services to meet the needs of sports tourists and players.

The hosting of the Olympic Games in Sydney in 2000 provided Australia with a once-in-a-lifetime opportunity. It enables Australia to showcase itself to the world as a tourism destination and a country capable of hosting such large-scale events. According to The Independent (2010), during the Sydney Olympics in 2000, the Australian Bureau of Statistics (ABS) reported that 4.94 million people visited Australia for a short period of time. This is an increase of 11%, or 480,000 people, over the previous record set in 1999. The increase was primarily due to a dramatic increase in tourist arrivals in December, which increased by 23% over the same month in 1999.

Sports had become an integral part of the Australian way of life. It is very important in Australian culture. This experience is a motivator for international visitors to come to Australia. According to an ABS survey (2005–2006), 66% of the population aged 15 and up (10.5 million people) participated as a player in one or more sports or physical recreation activities at least once in the 12 months. The participation rate was highest for those aged 25 to 34 years (75%) and then declined with age to 49% for those aged 65 and up.

Here are some of the marketing strategies used by Australia to promote sports tourism in order for marketers to successfully reach out to the target segment. It is critical for sports marketers to consider how each of the marketing mix elements is interconnected and related to the others. For instance, if sports products are not correctly priced, it will have a significant impact on how consumers perceive the organizations.

1.6.6. The Host Community and Its Role in Sport Tourism

Sports tourism is one of the fastest-growing segments of the tourism industry. Whether sports are the primary goal of the trip or not, an increasing number of visitors are involved in participating in sporting activities while on vacation. Sporting activities of different sizes and types draw visitors as participants or spectators, and destinations aim to differentiate themselves by adding local flavors to them in order to have authentic local experiences. Mega sporting events, such as the Olympics and World Cups, will serve as a catalyst for tourism growth if they are successfully leveraged in terms of destination branding, infrastructure development, and other economic and social benefits.

Individual sporting activities are on the rise, which is fueling the growth of sport-related vacations. Tourists seeking authentic experiences in host-guest interactions, physical health and well-being, and consumption of local goods and services exacerbate this. Communities, especially those in developing countries, have yet to fully realize the potential of local sports as a resource and competitive advantage in the development of their destination.

1.6.7. The Role of Sport Tourism in Employment, Income, and Economic Development

Sports tourism has a significant impact on the developing economy and can provide enormous opportunities for developing non-industrial regions. It is also the fastest-growing segment in the global travel industry, with annual revenues in the hundreds of billions of dollars. Sports tourism is a revolutionary phenomenon in the world that is experiencing the greatest growth in the tourism industry. Many countries have appropriate standing in terms of sport tourism and, as a result, contribute directly to their country's economic prosperity. Sports tourism is a relatively new and rapidly growing sector in the tourism industry, with an emphasis on developing-country target planning. Sports tourism is also expected to restructure rural and urban communities from a social and economic standpoint. Sports tourism appears to increase individuals' life satisfaction through tourist attraction and results in the economic well-being of local communities. Sport and tourism are now significant economic activities in both developed and developing countries (Swart and Bob, 2007). According to WTO reports, the tourism sector employed 43% of the world's workforce in 2010. Thus, the sport may have a significant impact on the functional and visual dimensions of sport-based facilities and goods. As a result, it is regarded as one of the most powerful factors influencing national development and revenue in the 21st century (Brown and Nagel, 2002).

Tourists are a major source of income and employment for the population of the hosting region (Batyk and Ski, 2009); this is why world metropolitans regard Olympic games as a unique opportunity for local marketing, since a potential economic advantage of hosting a big sports event is to attract more spectators and tourists to the hosting city, and therefore, it shall bring positive results (Preuss et al., 2007). The Olympic Games' long-term benefits have primarily centered on new infrastructures and facilities, city rehabilitation, international credibility, tourism development, public welfare improvement, job creation, and local employment opportunities. It goes without saying that this is why, in recent years, countries have chosen to host upcoming competitions (Kasimati, 2003). Then, sports tourism is viewed as an effective tool for combating poverty and unemployment, as well as increasing economic development and social welfare in communities, especially in developing countries.

Chapter 2

Sport Tourism Destination: Changes, Trends, and Management

CONTENTS

2.1. INTRODUCTION

The World Tourism Organization (WTO) defines tourism as any activity in which people travel outside of their usual neighborhood for recreation, business, or medical purposes for a period of no more than one year. Some tourism scholars define tourism as the temporary movement of people away from their natural environment, as well as the activities they engage in while in the new location, as well as the facilities built to meet the needs of tourists. According to experts, approximately 70% of tourists visit the 10 main tourism destinations, while 30% visit the other destinations. As a result, it is clear that the majority of the revenue produced in global travel goes to existing destinations; as a result, undeveloped destinations must engage in some strategic planning in order to draw visitors to their destinations. Tourists are attracted to a particular tourism destination because of its attraction, which represents the tourists' feelings about the destination's perceived ability to fulfill their needs. As a result, the more a particular destination meets the needs of tourists, the more appealing it is perceived to be, and hence its popularity increases. It is critical that the idea of sustainable tourism growth be implemented in all tourism destinations in order to adequately appeal to the needs of visitors and host destinations while also improving and preserving opportunities for the future.

It should be remembered that only about two-thirds of global travelers are estimated to travel solely for pleasure. The WTO has described many forms of tourism, including coastal and beach tourism, sports tourism and other water-related recreations, wildlife, mountain climbing, city tourism, and nature tourism. The aim of this chapter is to explore the tourism industry in light of the most visited destinations, as well as the social, cultural, and geographical features that draw people to those destinations.

2.2. TOURISM DESTINATIONS: AN OVERVIEW

Some scholars have proposed tourist destinations as a term with a definition, stating that destinations are places that have built an assortment of tourist products and services whose consumption is marketed under the destination's brand name. The destinations are geographically defined, and tourists recognize them as distinct entities with a number of core provisions (Zygmunt, 2013). Accessibility, facilities, service activities, and auxiliary services are examples of these provisions (Figure 2.1).

Figure 2.1. A destination, according to the World Trade Organization, is a distinct location where a visitor spends at least one night and exhibits tourism products such as attractions, support services, and tourism resources, along with defined management, physical, and administrative boundaries, and a very well image.

Source: https://th.bing.com/th/id/Rd7fa37ce440eb1ffbc48064d3d502594?rik=hp4KRK9ca9GBng&riu=http%3a%2f%2fwww.trbimg.com%2fimg-54a1b87b%2fturbine%2fla-tr-2015-travel-destinations-20141229-001&ehk=lBjfNa5vdTrEjRlZ8sIOTg1LEBzEpfq0eNqvC%2fYTb9E%3d&risl=&pid=ImgRaw.

Around 70% of tourists are estimated to visit the 10 major tourist destinations, while 30% visit the other destinations. As a result, it reflects the level of competition for less well-known destinations (Biganoa et al., 2004). The World Trade Organization divides tourist destinations into several world regions: the Americas, Europe, the Pacific, East Asia, the Middle East, Africa, and South Asia. However, the American region can be divided into North America and South America since these two blocks have very distinct but different, socio-cultural, ecological, and economic orientations (Vengesayi, 2003). South America, for example, is known for its biodiversity, whereas North America is not.

2.2.1. Attractiveness and Competitiveness

Tourism development holds an indispensable place in policy formulation in all countries due to the benefits it provides to the local people in terms of social, economic, and environmental concerns. Tourist destinations compete for tourists in the tourism market, resulting in a fierce competitive battle (Zygmunt, 2013). The main goal of regional and national governments, as well as tourism industry stakeholders, is to succeed by increasing the competitiveness of their destination. Because the tourism industry is the largest player in most countries' economic sectors, it is an important factor in the development of any region, particularly Europe. Because of its multiplier effect, it contributes to the stability of both regional and local economies, as well as affecting the creation of employment in the region (Vengesayi, 2003).

Tourists are drawn to a particular tourism destination because of its attractiveness, which reflects the tourists' feelings about the destination's perceived ability to meet their needs. As a result, the more a particular destination meets the needs of tourists, the more appealing it is perceived to be, and thus its popularity grows. The characteristics of the destination or the factors that make the given destination capable of meeting tourist needs are included in the ability to meet tourist needs (Biganoa et al., 2004). As a result, people assess the attraction of a destination and make decisions accordingly. As a result, the attraction of a destination is the primary factor in attracting tourists to visit and spend time at a given location. As a result, the value of a destination is determined by the attraction it has on visitors. If a destination lacks attractiveness, it will not attract tourists, and thus there is no need to build tourist facilities and services.

To be competitive in the tourism industry, both short and long term, operation, and implementation must be guided by competitiveness as just a new paradigm. Competitiveness is now recognized as a critical factor in influencing the competitiveness of industries, organizations, and even countries. A destination is competitive if it has an increasing market share as measured by visitor numbers and financial returns (Vengesayi, 2003). Thus, competitiveness is associated with a wide range of visitors and rising revenue. However, it is critical to examine the tourism sector through the lens of the industry's multifaceted nature, as there are numerous players involved in making the destination appealing and competitive.

2.2.2. Tourism Destinations Trends

According to the United Nations WTO, the gross revenue from foreign tourism in 2013 was one trillion and a half dollars. In 2013, the tourism industry received $1,049 billion from foreign tourists' spending on food, lodging, beverages, entertainment, and other related services worldwide (UNWTO, 2014). Taking inflation and exchange rate fluctuations into account, the industry's growth outpaced the previous trend in real terms. The increase in sales corresponded to a 5% increase in international tourist arrivals. International tourism accounts for an estimated 29% of global service exports and 6% of total exports of both goods and services (UNWTO, 2014).

According to the UNWTO (2014) report, Asia, and the Pacific are the fastest growing tourist destinations, while Europe has the largest share. In 2013, Europe accounted for 42% of global international tourism, while Asia and the Pacific accounted for 31% (UNWTO, 2014). The Americas accounted for 20% of total receipts. The United States of America came first in the list of most visited destinations, followed by Spain.

In recent years, the BRIC countries of Brazil, Russia, and China have created the fastest rising outbound markets. In 2013, the BRIC countries invested 40 billion dollars, compared to a rise of 81 billion dollars in overall tourism spending (UNWTO, 2014). China has led the world in both growth and investment over the last 2 years. Among the 10 most popular tourist destinations, Asian destinations, the state of Thailand, Hong Kong, and Macao saw some exponential gains consistently getting 18%, whereas Britain and the United States of America, at 11% recorded double-digit growth. Spain, France, Italy, China, and Germany all experienced 1% to 5% rise (UNWTO, 2014). Europe appears to have the fastest growth rate in terms of tourist income receipts. However, Asia, and Pacific destinations saw strong growth in terms of both arrivals and receipts (UNWTO, 2014). Despite its vast natural resources in forests, beaches, game parks, and reserves, the Africa destination has seen little growth and continues to lag behind. The continent's cultural diversity is still an underutilized resource.

2.3. TOURISM DESTINATION: THE NETWORKING APPROACH

Tourism destinations are "the fundamental units of analysis in tourism" (WTO, 2002). In its origin, the term 'tourism destination' can be stated as a typical geographical term and is understood as an element of geographical space. This approach is visible in the classic definition by Burkart and Medlik (1974, p. 46): "tourism destination is a geographical unit visited by tourists being a self-contained center." At present, even though it is one of the most commonly-used terms when analyzing tourism phenomena, one cannot state that there exists a single, generally accepted definition or even approach to this term. As the subject of analysis of many different sciences, including human, social, and life sciences, it started to be understood in many different ways. It should be no surprise that the approaches developed by economists, sociologists, regional, and physical geographers, social geographers, and others differ. Furthermore, as models and approaches developed by specific sciences become more sophisticated, they make it more difficult for specialists to achieve their research goals while also making it more difficult for researchers from different sciences to understand each other. This section seeks to present an approach developed on the nexus of economic geography and economic sciences, particularly management and new institutional economics. The networking approach to tourism destination research is still regarded as a promising method of comprehending the term. However, for a variety of reasons, the rules and tools developed by network analysis cannot be easily shifted from management theory to the evaluation of tourism destinations. The presentation of those reasons, at least those that were chosen and postulated to be the most important, generates content for the paper's discussion section. Despite the fact that this section focuses on an approach that is well-suited for economic geography and economic analysis, it should be noted that the tourist destination is still a multidisciplinary issue. The analysis and conclusions will be unbalanced if a multidisciplinary approach is not used to address this topic. The demand side approach, which is difficult to incorporate into the network approach, is extensively presented here as an attempt to avoid this type of imbalance.

2.3.1. The Definition of the Tourism Destination Concept

2.3.1.1. Classic Spatial Approaches

One of the most influential definitions of a tourism destination is the one given by Goeldner and Ritchie (2003, p. 466) in their world-wide recognized textbook, which states that "tourism destination is a particular geographic region within which the visitor enjoys various types of travel experiences." Other definitions that underline the spatial nature of tourism destinations are those by Murphy (1985, p. 7). One of the most detailed definitions within this classic approach is the one given by Framke (2001, p. 5), which states that "tourism destination is a geographical area, which contains landscape and cultural characteristics and which is in the position to offer a tourism product, which means a broad wave of facilities in transport-accommodation-food and at least one outstanding activity or experience." Finally, the definition given by Seaton and Benett (1997, p. 351), who were focused not only on the physical features of the place but also on intangible characteristics, is worth underlining. The last two definitions open new opportunities and a new approach which is much more connected with the achievements of economic geography and/or economics.

2.3.1.2. Economic Geography and Economics Approaches

Following the achievements of the economic sciences, economic geographers frequently perceive a tourism destination not only as a part of geographical space, but also as an important component of the tourism market, which can be described by characteristics of tourism demand and features of tourism supply. Similarly, two major attitudes can be identified within economic approaches to tourism destination analysis (Ewing and Haider, 2000, p. 56). The supply side approach is developed by both economists and economic geographers, whereas the demand side approach is more common in the economic sciences. Specific tourism destinations are still perceived subjectively in approaches typical of demand side analysis. According to Hu and Rithie (1993, p. 25), a tourism destination "reflects an individual's feelings, beliefs, and opinions on destinations and sees the ability to ensure satisfaction with his holiday special needs." This approach focuses on specific (past, present, and future) tourist perceptions and market choices. The term "destination" refers to a place or region where tourists choose to visit (Flagestad, 2002, p. 3). In this perspective, a tourism destination can be viewed as a perceptual concept that consumers interpret subjectively based on their travel itinerary, ethnic background, significance of visit, educational level, and previous experience (Buhalis, 2000). This leads to the conclusion that a destination is not only something that exists-it is also something that is imagined to exist, a psychological concept in the minds of its tourists and potential tourists (Seaton and Bennett, 1997). This approach, which focuses on the consumer and his or her perceptions, is a pillar of modern tourism marketing. Tourism destinations are frequently viewed as brands available to tourists and as alternatives in their choices (Konecnik and Ruzzier, 2006, p. 2; Hosany, Ekinci, and Uysal, 2007; Dawes, Romaniuk, and Mansfield, 2008). Destinations have considered as the leading brands in the travel industry, according to many researchers (Morgan, Pritchard, and Pride, 2002; Pike and Page, 2014). Woodside developed the theory of destination choice based on a division of available destinations into decision sets-similar to the process seen in other markets-as early as the 1970s (Woodside and Sherrell, 1977; Woodside and Lysonski, 1989). In this sense, specific tourism destinations compete for the attention of a tourist. The metaphor of a tourism destination as a brand enabled researchers to incorporate many achievements of corporate marketing and management into tourism destination practice (Hankinson, 2004; Kozak, 2004; Ema, 2010a). This subjective perception of a destination's boundaries corresponds to a social consciousness search for definitions of a region (of any kind, not just touristic) (Paasi, 2001; Chrom, Kuerová, and Kuera, 2009; Semian, 2012). This approach is gaining popularity among social geographers. The demand side approach, which is also very useful in marketing analysis and strategies, as well as in describing competition among destinations, has some limitations. Subjectivities in perception of specific destinations and their boundaries make analysis and management processes extremely difficult. One of the most significant contemporary challenges for destination marketers is the incoherence between the demand side strategy and the supply side strategy, which is focused on internal processes. Marketing actions are typically funded by a single entity or a group of entities located within some kind of boundary, particularly administrative boundaries, as public administrations are frequently involved. Tourists, on the other hand, often do not perceive administrative boundaries because they have their own, generally subjective, image of the destination they decided to visit. As a result, tourists may gain a leaflet promoting an administrative region without knowing where it is located. This is exactly the case when German tourists are given brochures inviting them to polish administrative regions, i.e., voivodeships, at ITB (Internationale Tourismus-Börse) fairs. When defining a tourism destination from the supply side, a different point of view is accepted. In this context, a tourism destination is most commonly defined as a location with a high concentration of tourism demand and supply, as well as the consequences of these factors, which include economic, social, environmental, and other factors. As a result, physical geographers and even sociologists can benefit from this approach. The supply-side approach, on the other hand, is not uniform. The fundamental definitions are centered on the examination of specific phenomena visible in tourist destinations. This is frequently the development of tourism companies and infrastructure, as defined by D'Angella and Sainaghi (2004, p. 38), who define a tourism destination as "a geographic region where there is a concentration of small/medium-sized companies sharing a homogeneous background." Bordas (1994, p. 3), for example, defines a tourism destination as "a group of tourist attractions, infrastructure, equipment, services, and organizations concentrated in a limited geographical area." Elmazi, Pjero, and Bazini (2006, p. 2) provide a more complex definition. According to them, a destination "represents a spatial unity of the tourism offer, possessing the appropriate elements of the offer, being market-oriented as well as tourist-oriented, existing independently of administrative boundaries, and necessitating management." It provides the fundamental institutional framework for developing a tourism development concept in which the focus shifts from the accommodation facility to the entire surrounding region and its economic structure (town, region, zone, country)." In the supply side approach, a tourism destination is frequently perceived through the lens of its products. Destination is defined here as an "area that consists of all services and offers that a tourist consumes during his/her stay" (Bieger, 1998, p. 7) or an "amalgamation of tourism products that offers consumers an integrated experience" (Buhalis, 2000, p. 97). Other researchers (Seaton and Bennett, 1997, p. 351) emphasize the importance of a destination not only as a location where tourism products are offered, but also as the central tourism product that motivates all other products. It is also unclear whether a destination should be perceived as a single important product on the tourism market or as a collection of products available locally. According to Seaton and Bennett (1997, p. 351), a tourism destination "is one product but also many," emphasizing the concept's dual nature. Experiencing a destination as a product, i.e., an offer for tourists to spend their time, is much more in line with the demand side approach because it returns to customer perceptions and competing for what they choose. On the contrary, viewing a destination as a collection of products is more in line with the supply side approach. This reflects the fact that a destination's product may be targeted to multiple segments at the same time, offering various ways to spend time in the same location. Local offers for active tourists, culture lovers, or spa and wellness lovers, for example, may be perceived as different products of a destination, and the task for destination managers is then to manage the product portfolio. This type of portfolio management, however, differs from that used in corporations in that specific products cannot be considered separately. Tourist motivations are much more complex than simply participating in one type of tourism, and they may engage in other activities during their stay at a destination in addition to the activity that is their primary motivator. Furthermore, some local offers may be shared by participants in various forms of tourism, implying that specific products share components. Accommodation facilities used by different tourists are frequently examples of such a common component.

The views expressed above are typical of economists and/or economic geographers. However, within the economic sciences, the concept of tourism destination has also become a topic of study for management students. A tourism destination is the most important unit of tourism management applications (D'Angella and Go, 2009). Typically, management science researchers define a tourism destination in a supply side approach, but the emphasis is on management processes and structures. This approach is more complex, and a portion of it is still controversial, because tourism destinations cannot be considered formal organizations, and there are no formal hierarchical structures in place. This is because a destination is typically made up of a number of individual businesses that offer "their" product in a disjointed manner (Flegestad, 2002, p. 3). However, as Capone and Boix (2003, p. 2) point out, the chaotic, uncoordinated development of tourism supply can be replaced by the cooperative activities of specific entities. A destination is then viewed as "a collective producer in a firm-like structure coordinating complementary services based on the needs and preferences of the target market segmented and marketed as one unit under one brand" (Flagestad, 2002, p. 3). As a result, Pechlaner (1999, p. 336) defines it as a "process-oriented unit of competition that must be capable of providing products and offers for defined target groups and guest segments." The metaphoric presentation of a destination as an entity equivalent to a company was needed in order to incorporate a rich variety of tools "borrowed" from the much more developed area of corporate management. However, it soon became clear that, while destinations must compete in the tourism market in a manner similar to how companies compete, as specific market entities that are not even a formal organization, destinations have so many and so strong distinguishing features that a simple implementation was not required and sound adjustment was required. The adjustment included, among other things, viewing a destination as an entity similar to a strategic business unit (SBU) of a diversified company rather than as a company itself. As an SBU in a diversified company, a place (region, city, country, etc.), is also diverse in its activities and products offered on internal and external markets, and tourism is just one of those activities/products. This way of thinking can be found in Pechlaner's definition, but it can also be found in the works of Gnoth (2004); and Bieger (1998).

2.3.2. The Network Approach to Tourism Destinations

More sophisticated approaches, particularly the systems and networking approach, could have been built with further development and joint use of spatial and economic supply side definitions of tourism destinations. In the systems perspective, a destination is defined as a non-administratively constrained area where tourist aspects are interconnected and system-integrated. This has consequences for travel motivations, trips, and the industry mechanism. This system, according to Elmazi, Pjero, and Bazini (2006, p. 2), includes the following subsystems: public self-government systems, entrepreneurial systems, and other systems; however, this view may be oversimplified, as the number of subsystems and their relationships may be greater. The evolution of the systems approach, which analyzes the complexities of tourism destinations, has created new opportunities for constructing a modern network approach to destinations. This approach was made possible by a kind of evolution that occurred in the main sciences developed to tourism destination analysis at the turn of the century. In sociology, management studies, economics, and regional and economic geography, the network approach is gaining traction. One of the most distinguishing characteristics of tourism destinations is the lack of hierarchical ties between the different organizations that offer products independently. This is why contemporary researchers prefer to talk about effective coordination and/or policy making rather than management. Since market structures and the hypotheses that followed them were further developed, the evolution and market effectiveness of destinations mostly on basis of management theories could be better understood. In today's world, numerous definitions are being used to define the boundaries of businesses. Companies began to look for sources of competitive advantage in non-competitive relationships with other entities, including competitors, according to a new paradigm in strategic management based on inter-organizational relations (IR). The networking theory (NT), which originated in sociology, began to be used in management studies (Provan, Fish, and Sydow, 2007). This theory was also found to be beneficial in better describing and comprehending processes that occur in tourist destinations. The networking approach is increasingly being used in contemporary tourism destination research. A 'network is a set of components, which we will recognize vertices and sometimes nodes, with links between them, called edges,' according to a simple and general definition (Newman, 2003, p. 167). Hall (2005, p. 179) defines a network in the business context as "an arrangement of interorganizational cooperation and collaboration." However, many different approaches and methods exist within NT that could and were introduced into tourism destination analysis, resulting in significant difficulties in developing a coherent theory of destination networking (Van der Zee and Vanneste, 2015). The theoretical literature on IR is fragmented, with contributions from various disciplines. Tourism researchers who attempt to incorporate it into tourism destination research identify several theories or micro theories that may or may not be valid. The most frequently cited theories are transaction costs, resource dependence theory, and NT. When analyzing the tourism market, the last option appears to be the most promising, as tourism can be described as a network industry par excellence (Scott, Cooper, and Baggio, 2008). This claim is supported by the definition of tourism as a system in which interdependence is essential, and collaboration as well as cooperation between various organizations within a tourism destination create the tourism product.

In this way, local alliances, agreements, and other formal and informal effective governance serve as a substitute for a tourism destination's fragmented nature (Scott, Cooper, and Baggio, 2008). NT has been proposed as a method for better understanding ongoing marketing activities and processes aimed at business development (von Friedrichs, 2007). According to Buhalis (2000), most destinations are made up of networks of tourism suppliers, and the advantages of such networks involve a more profitable tourism destination. Within NT, a tourism destination can be thought of as a group of interconnected stakeholders integrated in a social network (Scott, Cooper, and Baggio, 2008). Such a framework of stakeholders interacts and collaborates to meet the needs of visitors and create the experience that they consume. Accommodation establishments, tourist attractions, tour companies, as well as other commercial service providers are among the destination stakeholders, as are government agencies and tourism offices, as well as representatives from the local community. Interaction among these stakeholders is dynamic, complex, and susceptible to external shocks. The basic assumption of tourism destination management is that the effectiveness of these joint activities can be enhanced through cooperative planning and organizational activities to the profit of individual stakeholders (Baggio, Scott, Cooper, 2010a). One reason for studying networks as a central component of tourism is that they serve as a foundation for collective action. Many of a tourism destination's main resources are "owned" by the community and are used collaboratively to attract tourists. These can be physical resources like beaches, lakes, scenic views, and national parks; built resources like museums, art galleries, and heritage buildings; or intangible resources like destination brands or the popularity of friendly locals. Such collective action does not always necessitate the formation of a network organization, but when resources are scarce and tourism decisions are not always seen as falling within the purview of the government, the response is frequently a network of the stakeholders involved (Scott, Cooper, and Baggio, 2008). Furthermore, networks are proposed to function as systems capable of organizing and integrating tourism destinations, thereby benefiting the firms involved, improving destination performance and quality, and stimulating the provision of 'wholesome and memorable experiences' for tourists (Zach and Racherla, 2011, p. 98). At the same time that the networking approach was discovered to be useful for business practice and studies, another process occurred. This procedure improved the networking perspective of tourism destinations. A new network approach to 'place' was developed within human geography. Nicholls (2009) distinguishes between two approaches: relational and territorial conceptions of place. Both investigate social relations in distinct locations, but they focus on the effects of these relations: the former examines the structured cohesion of relations in specific sites, whereas the latter focuses on the contingent relations of diverse (in terms of sociology and geography) actors. Amin and Thrift (2002, p. 72) conclude that places are locations where actors with different statuses, geographical ties, and mobility's interact in fleeting and unstructured ways, supporting the relational conception of place. In this perspective, place has special traits that influence social networks that evolve within it.

On the one hand, the proximity and stability associated with a specific location foster favorable conditions for strong-tie relationships. Furthermore, a particular place is made up of a number of contact points where various actors can interact with one another on a regular basis (Nicholls, 2009, p. 91). Works that analyze destinations as clusters may be mentioned as among the earliest attempts to present a tourism destination from a network perspective (Nordin, 2003; Weiermair and Steinhauser, 2003; Hawkins, 2004; Jackson and Murphy, 2006). Industry clusters exist when firms and organizations are loosely geographically concentrated or when a group of firms and organizations are involved in a value chain producing goods and services, and they are innovative (Enright and Roberts, 2001, p. 66). Initially, the advantages of industrial agglomeration stemmed from natural resources, spatial costs of external transactions (Scott, 1983), transportation organization and costs (Scott, 1986, Scott and Storper, 2003), labor, or economies of scale (Enright, 2003). Clusters, as per Porter (1990), are geographic concentrations of interconnected firms, specialized suppliers, service providers, companies in related industries, and linked institutions (e.g., standards agencies, universities, trade associations) in a specific field that compete but also collaborate. Porter's (1990) view that underpinning competition and cooperation among businesses within a cluster and searching for competitive edge in an economy in inventions and an extremely rapid search for development gave the cluster concept new impetus. His cluster theory has become the industry standard, and policymakers around the world have embraced Porter's cluster model as a tool for promoting national, regional, and local competitiveness, innovation, and growth (Martin and Sunley, 2003, p. 5). Regional clustering, according to many authors, is part of the new industrial order (Hospers and Beugelsijk, 2002; Marková, 2014) and can be perceived as part of a subnational or global innovation and production system (Guinet, 1999). At the moment, the cluster concept involved in knowledge transfer (Maskell, 2001), as well as linkages and interdependencies among value chain actors. It goes beyond the traditional notions of clusters, which included horizontal networks of firms operating in the same industry group on the same end product market (Enright and Roberts, 2001). Many comparisons can be seen between the contexts in which destinations and clusters operate, especially when accepting modern definitions of clusters that emphasize company cooperation (Nordin, 2003).

The role of public authorities is acknowledged in both cases. However, for tourism destinations, this function extends beyond simply creating favorable conditions for business development, as the public sector is also responsible for providing many critical components of a tourism destination product. The most significant difference is noticed in the sequential nature of product creation in industrial clusters such as Silicon Valley, the Italian fashion cluster, and the Swedish forestry cluster. A cluster is typically formed by a network of suppliers and industrial customers with a visible single company (or together with many similar competing companies) in charge of a final product and its sale to the final customer. Even taking into account that in a cluster there are usually many different products produced and that almost all of them are provided by a different type of company, this is not comparable to what can be noticed in a tourism destination: all network members produce only a portion of the service potential that covers a wide range of offers, and the final customers, i.e., visitors, benefit from a diverse range of offers.

The too simple implementation of the cluster concept into tourism research was also criticized as being too business oriented and ignoring the fact that cluster members-like tourism companies-are typically unable to generate the reasons for tourists to visit on their own. This role is played primarily by tourism goods, which are frequently free, and cluster implementation may result in the marginalization of their impact (Hassan, 2000). This legal protection of free goods, which are elements of comparative advantage and are understood in this way by Porter (1990) and his followers (Hill and Brennan, 2000; Nordin, 2003; Tallman, Jenkins, Henry, and Pinch, 2004) as less important in making an economy competitive, highlights the importance of implementing geographic and spatial approaches (Scott and Storper, 2003; Scott and Garofoli, 2007; Asheim, Coo, and Pinch, 2004).

As a result, conventional business-oriented approaches based on Porter's theory are insufficient to explain the phenomenon of clustering tourism destinations. Several researchers go over the differences and similarities between industry clusters and tourism destinations in depth. The industrial district is another concept developed in regional economics and economic geography and implemented to tourism destination analysis. Marshall (1898), who was attempting to explain the geographical concentration of English industries such as pottery, cutlery, and basket making, pioneered industrial district theory in the late 19th century. The theory of industrial districts was then applied to an area of Italy known as the "third Italy" in the late 1970s. These regions appeared to be rising exponentially than the rest of the nation and emerging from recessions with greater success. From that point on, the concept remains especially popular among Italian scientists (Becattini, 2002a, b; Corò and Grandinetti, 2001; Sforzi, 1989). Mottiar and Ryan (2006) define industrial districts as having a geographical and sectorial concentration of firms, small size companies, strong interfirm relations (IR), a social or professional milieu, and a focus on innovation. Similarly, Hjalager (2000) sees industrial clusters as having the following main characteristics: firm interdependence, flexible firm boundaries, cooperative competition, trust in long-term collaboration, and a "community culture" with supportive public policies. Cor and Grandinetti (2001, p. 189) provide a very simple definition of an industrial district, stating that it is a network of small and medium-sized enterprises (SMEs) embedded in a local context, which draws attention to the network shaped nature of industrial districts. All of these statements demonstrate that the concept of industrial districts contracts with similar phenomena as clusters, and that the concept, like the cluster, could be used to analyze tourism destinations. Tourism destinations, according to Hjalager (2000); and Mottiar and Ryan (2006), can be viewed as examples of industrial districts. This concept, after all, is not as prevalent as clusters among tourism researchers. This could be due to the core theory of industrial districts' emphasis on production sectors (Mottiar and Ryan, 2006). Also, the parallels between tourism destinations and industrial districts are less obvious, particularly in terms of governance structures. This is also true for the increased vertical division of labor between areas that provide tourist services and areas that do not (Hjalager, 2000). Another attempt to apply the NT approach to tourism destination analysis is Gnoth's metaphor of virtual service company, which can be defined as a network of enterprises that share resources and organize their cooperation as a collaborative effort (Gnoth, 2004). However, as Gnoth (2004) points out, there are significant differences between the typical virtual firms found in industrial markets and tourism destinations. For starters, in most tourism destinations, there is no single company in charge of overseeing the entire production process. Second, the contribution of each SME in tourism is not cumulative, as it is in the case of different companies in the automobile industry. Tourism is experienced rather holistically, and customer value is frequently created between the different services, as a blend of those services and the physical and intellectual assets of a given destination (Gnoth, 2004). Probably the most difficult proposition of how to evaluate tourism destinations in the approach of NT is the one offered by Scott, Cooper, and Baggio (2008). Complex quantitative methods are implied here in order to better understand the relationships between specific stakeholders and their impact on the overall network's effectiveness. NT is currently most commonly used to improve understanding of governance in tourism destinations. Governance could be defined as "the self-organizing interorganizational networks characterized by interdependence, resource exchange, rules of the game, and autonomy from the state" in the network approach, as opposed to the corporate approach (Ruhanen, Scott, Ritchie, and Tkaczynski, 2010; Rhodes, 1997, p. 15). Governance, on the other hand, is a concept that refers to the relationships between various stakeholders and how they interact with one another. It is concerned with how stakeholders determine, implement, and evaluate interaction rules (Baggio, Scott, and Cooper, 2010a, p. 51). According to Beritelli, Bieger, and Leasser (2007, p. 96), the concept of governance applied to tourist destinations entails establishing and developing policy rules and mechanisms, as well as business strategies, by implicating all institutions and individuals. Similarly, Nordin and Svensson (2007) concentrate on social networks and relationships, particularly those that exist between the public and private sectors. Because both the public and private sectors are involved, the governance dimensions used may be derived from those used in both sectors (Ruhanen, Scott, Rithie, and Tkaczynski, 2010, p. 5). The whole concept of destination governance is aimed at making the groups of organizations that combine together to form a destination context (Nordin and Svensson, 2007).

2.3.3. Special Challenges in Implementing NT in the Area of Tourism Destinations

Researchers frequently perceive a tourism destination as a network of stakeholders. The representation of a destination in the frame of NT facilitates a better understanding of the processes and phenomena that may be observed in destinations. The implementation of NT in tourism, on the other hand, is fraught with pitfalls and difficulties. Among them, spatial embeddedness, and the non-voluntary nature of membership in a network, the critical function of free goods in product development, the primacy of SMEs in a destination network, distinctions between particular destinations and the difficulty in clearly defining borders among networks, will be discussed here as the most critical items. Each of them could become more important if the methods and tools of analysis are directly implemented from other markets, particularly industrial markets (Ema, 2010b).

2.3.3.1. Spatial Embeddedness and the Non-Voluntary Nature of Membership in a Network

One of the most important questions that a company must answer regarding NT is whether or not to join the network (Möller and Svahn, 2003). In NT, a company is normally free to choose which network it wants to join or whether it wants to join any network at all. Destination tourism businesses do not have this option. They are interconnected with other interested parties in the destination product preparation, despite of whether they are willing to collaborate in the network. The rules of cooperation are drastically altered as a result of this spatial embedding. According to NT, a company that is dissatisfied with the results of its network membership may leave and seek other partners. This is a difficult decision to make but it is one that must be made. Tourism companies are "condemned" to coexist in a destination. A company cannot "escape" from its partners if they are hostile or irresponsible, and it cannot "escape" from a network if it is poorly organized and managed. Even if a company does not consider itself a network member and does not actively collaborate with other entities, it is influenced by network actions and the actions of other companies, and the network is influenced by this company.

2.3.3.2. The Crucial Role of Free Goods in Product Creation

One of the most difficult challenges in understanding network relations in tourism destinations is that the tourist experience is developed from the 'between' of services provided by specific local businesses rather than directly from those services. The major elements for visitors are frequently the location's natural or cultural resources, which are still available for free. The very first issue raised by this statement is the role of the public sector in the creation of tourism products (Flagestad, 2002). Future research should include answers to the question of what the consequences are of the fact that the external resources used by tourism enterprises are rather free resources which can be used concurrently by many firms, whereas resource dependency theory emphasizes mostly the possibility of using external resources that relate to other companies, particularly those that are not available. Albrecht (2013) discovered that while significant progress has been made in the investigation of private-sector networks at the destination levels, research on networks involving public sector stakeholders and networks along all sectors and levels of governance remains inadequate. Developing a research solution that enables one to consider the public sector as a holder of free resources, as well as the relationships between local actors and free resources, appears to be one of the most important tasks for tourism researchers looking for ways to implement the NT in the field of tourism destinations. In the destination context, free goods are not only tangible goods, but also marketing assets. The tourism destination brand and its attractiveness are two of the most important factors influencing the success of local businesses. In that case, the issue of so-called free riders, i.e., stakeholders who benefit from the efforts of other stakeholders without exerting their own, remains critical (Zmylony, 2009).

2.3.3.3. The Predominance of SMEs in a Destination Network

There is a lot of published evidence that tourism is primarily a SMEs industry (Go and Appelman, 2001; Woods and Deegan, 2003; Jones and Haven-Tang, 2005). Furthermore, much of the literature on the NT is concerned with large company collaboration, which is why the rules described there are not fully applicable to the SMEs-dominated tourism sector. The dominance of SMEs in tourism destinations has several consequences. On the one hand, the SMEs sector should be especially interested in collaborating because it may help them overcome market constraints caused by their small size and limited financial resources (Go and Appelman, 2001, p. 193). Because of its involvement in the destination network, a micro-firm, such as a family business in the lodging industry, maybe promoted globally. This is only one example of possible advantages. On the other hand, the industry's dominance by SMEs creates numerous difficulties and constraints in collaboration and network formation. The majority of them are entities that are or should be part of a network. The more companies that participate in a network, the more difficult its management becomes, especially when it comes to setting common goals for the network as a whole. Setting goals is made more difficult by the competitive relationships between SMEs and their various profiles (e.g., the various services and goods that companies offer visitors). Small-scale operations are frequently the cause of a dearth of intangible resources, particularly knowledge. As a result, there is a lack of professional management. There is no room for a manager position in small, family-run businesses where the founder also works at the front desk. In other terms, it is much more difficult to convince small business owners to begin collaboration because they may lack the necessary professional management knowledge to fully comprehend the benefits. SMEs frequently reflect the founder's personality (Keasey and Watson, 1993), and as a result, SMEs are frequently characterized by a strong will to survive. Running one's own small business is frequently regarded as the founder's way of achieving independence. This may result in the development of the so-called fortress mentality (Lynch, 2000), which is an impediment to cooperation (Simpson and Bretherton, 2004, p. 112). SMEs are also much more vulnerable to bankruptcy, and new businesses are formed on a regular basis (Wanhill, 2000), so the list of companies that are part of the destination network changes frequently. This is a significant challenge in establishing long-term sustainable relationships between companies in a destination suggested by the NT.

2.3.3.4. Differences Between Particular Destination Types

It is difficult to determine common rules of how to implement any competitive edge framework in the tourism destinations sector as destinations vary greatly from one another. Instead, look for concepts that are appropriate for specific types of destinations. The statement about SMEs predominating would be irrelevant in some destinations, such as large cities. Instead, the issue of collaboration between hotels that are part of global hotel chains may arise. Different destinations offer various products and are present in various markets. As a result, it is very likely that ideas that work in one location will not work in another. When implemented, different destinations may necessitate different modifications to the NT. One of the most common differences between destinations is the concept that examines the community and corporate models of destinations (Flagestad and Hope, 2001; Beritelli, Bieger, and Leasser, 2007). The community model represents a situation that is prevalent in most European tourism destinations, where multiple small-sized, mostly local businesses are associated with the development of tourism products. Furthermore, the role and support of the local government are relatively significant. The corporate model, on the other hand, is more akin to a North American destination style model, with unquestionable leadership of large, often external corporations in destination development. The management structure here is more like that of a company and is more integrated, hierarchical, and centralized than in community model destinations (Flagestad, 2002). Beritelli, Bieger, and Leasser (2007, p. 97) contend that the NT is the appropriate approach for community model destinations, whereas the dyadic resource dependency theory correctly explains the relationships within corporate model destinations. However, it may be questionable if, even in large company-dominated destinations where other companies serve as a leader's satellites, relationships between those satellites do not exist or do not play a key role in destination competitiveness. Even if this is accepted, the dyadic relationships between the focal company and individual satellite firms are likely to be mutually interconnected (Wang and Xiang, 2007, p. 76). Then, rather than a simple dyadic perspective, Jarillo's (1998) concept of a network within the hub firm would better fit the corporate model of destinations. However, this assertion does not change the fact that management in corporate model and community model destinations should use different NT approaches, as Jarillo's hub firm network does not match the community model. Finally, destinations that attract different tourism markets must deal with different features and customer behaviors, which may necessitate different methods of organizing the local tourism network. This may include long-haul versus short break holidays, for instance, or leisure versus business tourism destinations.

2.3.3.5. Difficulties in Setting Clear Borders Between Destinations

One of the key arguments advanced by proponents of IR is the difficulty in defining precise borders between companies, given outsourcing, resource dependencies, and cross-border activities. Instead, we should examine network competitive advantage. It is remarkable, then, that IR researchers may repeat the same error, but on a network level. In many cases, the whole network approach (Provan, Fish, and Sydow, 2007; McLeod, 2014), which is well suited for tourism destination analysis, results in a network of collaborating entities as well as a closed system separated from the environment, which acts on the market in a manner similar to Porter's (1980) value system. For at least three reasons, this approach is unsuitable for tourism destination analysis (Ema, 2014, p. 243). To begin, consider the hierarchical structure of tourism destinations. A single destination may be made up of several smaller destinations. A tourism region is made up of several communities, just as a country is made up of several regions, and so on. Different market actions necessitate different definitions of the destination. The most obvious reason for this is that visitors have different perspectives on tourism destinations, which is generally, but not always, a result of the distance between a visitor's residence and the destination. For locals, the destination may be a single town, whereas for tourists from other countries, the destination is usually the entire country. This ties together neighboring destinations in the same way that tourism companies are linked within a single destination (Bengtsson and Kock, 2000). Many actions on many markets necessitate the ability of the destination to cooperate with the neighbor in one market while competing with it in another. What strengthens this relationship is the fact that the destinations chosen from a supply-side perspective do not portray the perspectives of a specific visitor. While visitors stay in one location, they frequently visit neighboring areas, which means that having a neighboring destination with a highly competitive offer may not only be a threat, but also an opportunity for the destination. As a result, the importance of intra- and inter-destination collaboration is emphasized (Fyall, Garrod, and Wang, 2012; Ema, 2014). Second, because destinations are established using different criteria in setting the borders, many tourism entities, also including companies and localities, may participate in more than one destination at the same time. This complicates the issue of competition and cooperation among destinations. Finally, the product of a destination is established not only by internal entities, but also by external stakeholders, whose role must be acknowledged. Some of them, such as investors, may be regarded as inextricably linked with a destination; however, there are firms that collaborate with a destination's product creation and marketing only inadvertently. This could include tour operators as well as companies such as breweries, sports equipment manufacturers, and others that run joint promotional campaigns with the destination. They may collaborate with the destination as a whole in some cases, but they also collaborate with specific companies within one destination (Ema, 2010b). It can also be seen that the current section leads to the development of more difficult and comprehensive methods of interpreting the term. The networking approach is one of the most popular in the 21st century. Over the last two decades, an increasing number of studies on the role of networks in tourism have been published. Tourism destinations are a unique type of network, and collaboration processes are critical to their success. Once researchers and practitioners accept the preceding statement, they can begin looking for the best approach to creating long-term competitive advantage in the tourism market within the framework of NT theory. However, the distinctions between traditionally understood company networks and tourism destinations as presented here make NT difficult to implement and full of traps.

2.4. SPORT TOURISM DESTINATION ANALYSIS

The study of resources is a critical first step in the growth and management of sport tourism destinations. Different destinations have distinct sport tourism experiences since they have distinct sport and tourism infrastructure. Within the tourism sector, these resources can pertain to tourism infrastructure, levels of service growth, tourist attractions, and information services. Sports venues such as stadiums and arenas, educational facilities, sport science infrastructure, and sports medicine facilities are examples of sports resources.

Figure 2.2. The size of tourism growth varies greatly between local and regional destinations.

Source: https://assets-corporate.visitvictoria.com/images/_1144x545_crop_center-center/2018-Grand-Prix.jpg.

There is a significant difference between developed sport tourism opportunities and those that are largely dependent on nature. At many urban sport tourism destinations, the resource base for sport tourism is increasingly centered on major stadium facilities that have recently been planned and established in tourism precincts alongside other entertainment sites, attractions, and tourist services (Figure 2.2).

At the regional community level, the resource base for sport tourism generally includes local club-based sports facilities, as well as community-funded walkways, riverside bike paths, and municipal leisure facilities such as swimming pools and outdoor racquet sports venues. The possible benefits of sport tourism at the community level, on the other hand, are the same as it takes place in smaller geographical areas. Accessibility, distance from tourist-generating areas, and a lack of tourism infrastructure and service growth are common challenges for such destinations. Major challenges exist for sport tourism destinations situated in rural or peri-urban areas.

Dynamic hierarchies of existing sport tourism resources exist. This is true for both natural and man-made sport tourism opportunities. Nature-based sports provide excellent prospects for destinations that are well-positioned in the sports location hierarchy. Ski-jumping, surfing, hang gliding, and surf lifesaving are all closely connected with famous locations due to natural phenomena and/or climatic conditions that occur there. The advent of new or increased participation in existing nature-based sports such as kite-surfing and base-jumping presents significant opportunities for tourist destinations that (a) identify the natural resources that may serve new types of tourist demand and (b) react efficiently and effectively to emerging sport tourism opportunities.

The position hierarchy theory is similarly applicable to the construction of sports facilities. Baseball facilities in the United States vary from local parks around the country at the bottom of the ladder to venues such as Yankee Stadium (New York) and Camden Yards Stadium (Baltimore) at the top. The position hierarchy of cricket in England varies from small village cricket grounds to globally renowned locations such as Lords (London) and Edgbaston (Birmingham). Each clearly caters to very different athlete and spectator markets, with the difference between professional and amateur sports facilities being most apparent, and offers equally distinct visitor experiences. Sport tourism infrastructure and facilities located in the local hierarchy contribute to the resource base for sport tourism, but in somewhat different ways.

There are several approaches to comprehending sport tourism capital. This involves the categorization of resources that are either natural (though typically with some sort of cultural development) or designed specifically for the purposes of sport. The quality of sport tourism services, whether natural or man-made, will have an impact on the local/regional, national, and/or international tourist catchment areas that they can affect. Thus, an overview of sport tourism resources is incomplete unless the nature, uniqueness, and status of those resources, as well as their tourism market range, are taken into account.

The challenges of resource quality and market demand are extremely important for sport tourism destinations. The resource requirements of a given sport vary greatly depending on the form of athletic pursuit, with notable distinctions between amateur/professional, novice/elite, and recreational/competitive participants. The sport of skiing perfectly illustrates this point. This sport's physical resource requirements are basically skiable terrain and sufficient seasonal snow cover.

As part of the destination resource analysis method, Standeven and De Knop (2012) emphasize the value of understanding tourism infrastructure and superstructure. Accommodation, transportation facilities, visitor information systems, ancillary attractions, and tourist opportunities are all essential components of a destination's resource base. The proximity of sport tourism destinations to visitor markets and accessibility are significant considerations in the analysis of sport tourism destinations, especially for regular-season sport competitions and one-time or recurring sports events. It is worth noting, however, that for others, the experience of exotic and exclusive destinations, and maybe even travel to the destination itself, is an integral part of the overall tourist experience.

Sport tourism resource research must take into account factors other than the "bricks and mortar" of sports facilities and tourism infrastructure. Institutional structures are an important component of resource analysis. The willingness and ability of a destination's policy and planning agencies to cooperate is a critical political dimension of the sport tourism resource review. Coordinated public and private sector interests, as well as the availability of capital investment, are also important. These human and financial resources are critical in the field of sport tourism, assessing factors such as the timely and effective construction of sports facilities, as well as the success of national and international sports events at a destination. Cultural capital is one of the human resources important to sport tourism destinations. Sport tourism has been described as an experience of physical activity, either active or passive, that occurs within the context of a specific place's experience. As a result, the cultural text with which sports are associated in a given location is a significant component of sport tourism. Because of the prestige and importance of the events that they host, the Masters golf courses and grand slam tennis venues stand out from other venues. Wimbledon's legacy and aura of tradition elevate it above all other tennis sites, with consequences for tourist market catchments. Elements of uniqueness foster the status of sport tourism destinations, as well as the importance of images associated with sports that take place at specific and identifiable tourism destinations.

An awareness of sport and tourism resource management at a destination is incomplete without a recognition of sport tourism participants. Sport tourism is commonly regarded as a niche segment of the tourism industry that can be targeted in order to expand the suite of tourist audiences drawn to a destination. Participants in sport tourism are often regarded as a collective market that can be treated generically, as shown by the marketing of major sporting events to mainstream audiences. However, the fact is that sport tourism is made up of a wide variety of niche markets. During the run-up to the 2000 Olympic Games, for example, many Australian cities targeted particular consumer segments. As a result, while Bull and Weed acknowledge that 'sport tourism (represents) a series of separate niches,' the degree to which sport generates a diverse range of niche tourism markets is underappreciated. Understanding sport tourist markets is a critical step in determining a destination's sport tourism growth potential.

It is critical that destination studies take into account sport tourism industry demand. Sport tourism markets can be divided into leisure and business travel markets at the most basic level. Both are distinguished by a wide range of specialized markets that can be profiled for destination marketing and growth. Visitors who attend non-elite sporting activities as competitors or observers, activity holiday centers, dream camps, and outdoor adventure pursuits are examples of the former. The latter involve professional sports teams, team management, elite athletes, and sports marketing agencies, media, and professional sports administrators. Within these broad categories, there are a plethora of unique niche markets that can be found and, if strategically advantageous, targeted by sport tourism destinations. Then, destination managers must consider what motivates sport visitors and the experiences that members of particular market segments pursue. Tourist motivation research is concerned with why people travel, the benefits they seek, and the experiences they seek to fulfill their needs and desires. Tourist motivation is a feature of the traveler's self-perceived needs, which influence decision-making and the purchase of tourism items. The traveler's motivational profile is made up of both intrinsic and extrinsic influences. These factors have been classified as push (psychological) and pull (cultural). The former is intrinsic and special to each visitor, as they are defined by the individual's personality and attitude. Price, destination image, marketing, and promotion are all pull factors. Destination image, which is a feature of physical and abstract qualities, plays a significant role in the formulation of expectations. Attractions, events, sports facilities, and physical landscapes are examples of physical attributes. Atmosphere, crowding, protection, and ambience are examples of abstract qualities that are more difficult to quantify. Tourism companies may use these characteristics to cultivate a distinct destination brand and advantageous destination imagery.

Sport tourism, like other types of travel, involves a collection of motivations that are formed in anticipation of the fulfillment of desired needs. Sport tourists can, to some degree, be driven by push factors-including escape from daily life, the quest for camaraderie, and to establish friendships and a sense of belonging. Additionally, or similarly, sport tourists can be attracted by pull factors such as the atmosphere and excitement of professional sports. Tourist motives are important for understanding why people travel or do not travel, their destination preferences, and other aspects of tourist activity.

An understanding of the motivations and desired experiences of leisure travelers is key to the success of sport tourism destinations. If satisfactory visitor experiences are to be achieved, the relative importance of motives such as competitiveness, mastery of specific skills or disciplines, engagement, camaraderie, and experiencing the tourism product at a tourism destination, all of which may differ between specific travel markets, must be understood. The potentially informed analysis of sport tourist motives and perceptions represents a significant contribution to the study of sport tourism.

In contrast, the needs of business travelers are primarily focused on achieving professional business objectives. The infrastructure and service requirements of sport tourism-related businesses are a clear distinction between leisure and business sports travel markets. It is worth noting, however, that business travelers, whether professional athletes, team managers, or members of sports associations, are no less likely to have leisure and recreational desires that must be met at the tourist destination.

This part of sport tourism represents a significant avenue of investigation, given the increasing significance of professional sports as just a travel market and the specific motivating factors and needs of professional sportspeople. The team sports travel market in the United States is projected to be worth more than US $6.1 billion per year. Since professional sports competitions are competitive and changeable, options for tourist destinations to serve as a base for a professional sports franchise (or seasonal base for touring sports teams during times of competition), or to host elite level sports competitions, are continually renegotiated.

Understanding sport tourism destinations necessitates an understanding of why people are drawn to various locations. Any such comprehension necessitates knowledge of both supply (destination resources) and demand (visitor demands) for sport tourism. In terms of availability, it is important to consider what services for sport tourism are available at a destination, as well as what visitor experiences those resources provide. Similarly, an understanding of the values, desires, and motivations of established consumer groups, as well as the characteristics of members of specific sport-related tourism markets, is needed on the demand side. Such insights allow sport tourism practitioners and destination managers to fully appreciate sport tourism's growth potential. They can also allow for the identification and evaluation of emerging sports-related growth opportunities in a timely and competitive manner.

2.5. SPORT TOURISM DESTINATION RESOURCE ANALYSIS

The perception of location is an important component of the sport tourism experience. Not only does the distinct realm of tourism entail people being drawn to destinations with distinct characteristics and values, but sport often necessitates unique spatially located services. These services can require unique environments or facilities, but the key point is that they are not ubiquitous; they can only be found in specific locations. Although some services are more widely available than others, such as paths for running or cycling, facilities for activities such as skiing or rock climbing are less common. However, even when services are more readily accessible, the standard can vary, with high-quality resources only being found in a few places. Football in a premier league stadium is a very different environment to football in a local park, and cycling through scenically appealing environments of national parks compares sharply with cycling through busy streets in towns and cities. Thus, participation in sport (whether as a participant or spectator) will often necessitate travel, some of which will undoubtedly be to destinations other than one's self, and this will necessarily constitute sport tourism. In order to gain a better understanding of the sport tourism phenomenon, it is necessary to first determine why people are drawn to various destinations, which necessitates an understanding of the locations involved. Various variables, such as inherent physical characteristics and attributes, accessibility, related infrastructural arrangements, and degree of growth, must be considered when assessing the resources of such destinations. However, the analysis is far from easy because not only does sport tourism entail a wide range of different activities, each with its own set of resource requirements, but also because the respective participants and spectators can differ in terms of their goals, motivations, and values. As a result, another collection of variables that must be included in the analysis is how resources are viewed and valued culturally.

2.5.1. Physical Qualities and Characteristics

Various attempts have been made to categorize physical resources related to recreation, leisure, and tourism, and abstracts of such classifications, as well as others, can be found in a number of recent tourism geographies. The definition of a 'continuum spanning from biophysical resources to man-made facilities' is implicit in much of the debate underpinning such classifications, a concept of significant relevance for sports resources because it accommodates outdoor pursuits at one end of the spectrum and those facilities, mostly urban-based, that have been explicitly built for sport at the other.

Clawson et al. proposed one of the earliest examples of such a classification, and probably 'one of the most useful,' based on position and other characteristics including size, major usage, and degree of artificial growth. Areas were organized on a range of leisure opportunities, from user-oriented to intermediate to resource-based, under this scheme. User-oriented areas were those that were close to users, had limited space requirements, and often had artificial features; they included services such as urban parks, swimming pools, golf courses, and playgrounds, where the landscape elements were less significant. At the other end of the spectrum, property-based areas emphasized the nature of the physical resource, with vast land units involved and remoteness as a basic ingredient. This group included national parks, forests, upland, and wilderness areas that catered to activities like orienteering, canoeing, hiking, and rock climbing. Both spatially and in terms of activity, intermediate areas were found between the two extremes. Most sites were reasonably accessible, with most being within one- or two-hours' drive of potential users. This category included camping, picnicking, hiking, swimming, hunting, and fishing facilities.

Burton (1995) established five distinct resource characteristics for tourism resources: climatic; coastal; landscape and wildlife; historic; and cultural, entertainment, and man-made. Standeven and De Knop recognize that this refinement of Clawson's categories is especially important to sport tourism, particularly the climatic resource aspect. While the historic element does not seem to be instantly important, it does accommodate the growth of nostalgia sport tourism, and the addition of the climatic element supports both the need for ambient climate for certain activities as well as very precise (even extreme) climates for others, such as snow skiing, surfing, and various extreme sports. Skiing, for example, necessitates a specific form of upland climate with suitable physical conditions, such as snow cover and slopes. Snow cover is the primary determinant of season length, and related factors such as duration, earliness of first snow, quality of snow cover, and consistency of snow cover from year to year are all important in determining the quality, and thus performance, of a particular location.

The Clawson scheme has been chastised for its rather ambiguous language, which requires a somewhat narrow understanding of the term 'resources' and seems to neglect the fact that all leisure facilities must be user-oriented to some degree. Nonetheless, it retains some contemporary significance because it can be adapted to include specific types of sport, and it also begins to provide a solution to the distance issue associated with sport tourism. User-oriented services will not be included in sport tourism under this scheme because they are specifically local, mostly being used after school or work, with no substantial travel involved, though some might be used by visitors who are visiting the destination mainly for other reasons. The other two regions, on the other hand, will be interested in sport tourism because resource-based areas are associated with vacations and intermediate areas with day trips and weekend visits. Unfortunately, such a system cannot handle all types of athletic activity.

The fundamental issue with the Clawson model is that it does not address certain quality issues. The concept of challenge is central to sports. This challenge may be interpersonal, but it is more often than not accompanied by some type of natural or manmade resource. It is also pursued at various levels, including recreational, competitive, and elite. As a result, sport is associated with a variety of different environmental/resource requirements. The first relates to the resources required to carry out even the most basic activity. Local parks, for instance, provide open space for different types of recreational sport, with 'jumpers for goalposts' being the pinnacle of resource provision. Higher-quality resources, on the other hand, maybe preferred, if not required, where sport is undertaken at a competitive or elite level, and such resources are only found in certain locations. They could be specific natural or semi-natural resources, including mountains, rivers, lakes, waves, and forests, that are located where they are due to physical geography. These resources, which cater to sports such as skiing, climbing, canoeing, surfing, and orienteering, would clearly fit into the Clawson model in terms of both environmental and locational characteristics. Other high-quality sports resources, on the other hand, can be found in a variety of locations, mostly in major urban areas. These resources are distinguished by purpose-built features that are specifically designed for sport, and their spatial distribution is influenced by economic factors like market thresholds, social, and political considerations related to social policy, or urban regeneration related to tourism. These sports facilities have a user-oriented location for the people living in such centers, but the pattern is clearly very different for those who have little or no such facilities in their home town and choose to travel to such centers to participate.

2.5.2. Accessibility and Infrastructural Arrangements

One crucial element in determining the value of resources available is the ease with which potential users can travel to connect them and the duration to which they can be fulfilled in the destination. In traditional spatial analysis, accessibility is linked to distance and the efficiency of transportation networks. The resources that are most accessible to the greatest number of users are generally regarded as more valuable than those that are less so, and simple concepts such as distance decay and market accessibility aid in explaining the locational value of most spatial phenomena. Veal, for example, has shown that distance, travel time, and travel cost all limit access to sports facilities, and it is certainly true that much competitive sport is established through local and regional leagues, ensuring that both teams and supporters do not have to travel far in relation to 'away' matches. Furthermore, Bale has demonstrated the importance of central place concepts in the distribution of sports places, such that those 'capable of drawing on a regional catchment area located further apart than those catering for a district catchment, which, in turn, are sited at more distant intervals than those only capable of drawing on a local sphere of influence.' While these concepts are also important for understanding the distribution of many tourist resources and associated travel patterns, the overall situation must be qualified by the fact that, with tourism, travel is not only a requirement, but it may also be a key motive. Ease of access by minimizing distance traveled may not always be the most important consideration. This is not to say that people want to travel long distances, as travel statistics show the importance of tourism activity between neighboring regions and countries. However, getting away from one's home environment is important, but for some, the journey itself may be part of the appeal. Various studies have suggested that escapism is an important motive for tourism, and Weed and Bull (2004) suggest that this would also be true for sports tourists, who are likely to enjoy integral approach by pursuing their sport in a variety of interesting and uncommon places away from their home environment.

Nonetheless, despite the importance of travel, most studies of sport tourism emphasize the importance of accessibility. The various case studies, which Weed and Bull (2004) recently discussed, all demonstrate the importance of accessibility: The relatively short flight times from northern European cities clearly influence the location of sports training camps in the Mediterranean; the transport hub advantages of large towns and cities, where motorway networks and intercity train services converge, and where international and regional airports are located, make them convenient locations for a variety of sporting events; In relation to this example, the proximity of the Alps has resulted in a variety of travel options-road, rail, and air as well as opportunities for weekend trips as well as longer stays. The impact of accessibility is shown in relation to the travel characteristics of UK skiers, with France, the most accessible location, ruling the market. France accounts for 36% of all skiing holidays in the UK, followed by Austria and Italy. While other factors, like purpose-built resorts with substantial accommodation oriented to the mass market and the introduction of a large number of tour operators specialized in these resorts, have played a key role, the ease of access to French resorts does attract independent holidaymakers who can travel by train, car, or fly cheaply with low-cost carriers. Accessibility is especially important for day and weekend skiers, as is the case for many North American resorts.

Another important factor in determining the quality and significance of sport tourism destinations is the existence of various facilities that allow visitors to come to the destination and meet their needs, such as lodging, catering, and entertainment. The importance of cities as sport tourist destinations stems in part from their established importance for other forms of tourism, as they can provide accommodation, restaurants, bars, clubs, and numerous different forms of entertainment for sportspeople in addition to the immediate attractions of sport. Cities can not only accommodate traveling spectators, but part of the appeal of traveling away to watch sport may include experiencing the nightlife and other services available to mainstream tourists.

The modern ski resort's unique environment is another example of a sport tourism industry that embraces more than just the basic resource needed for the sport itself. Ski resorts mandate a collection of built features, including ski lifts, ski schools, equipment shops, accommodation, restaurants, car parking, and various leisure facilities, in addition to a sufficient natural environment. However, these facilities can differ in terms of both their level of development and the quality of the accommodation provided. Barker (1982), for example, identified differences between eastern and western Alpine resorts, with the former (developed at lower altitudes) being more incorporated with the economic and cultural lives of indigenous communities than the high-altitude resorts of the western Alps, which involved more external capital and labor. Some resorts have also created a tremendous summer season based on warm-weather pursuits or developed means to transport tourists to high altitude skiing areas, and Shaw and Williams (2002) have created a continuum of skiing resorts based on these differences, varying from those at the low altitude end of the continuum with specialized activities to those at high altitudes establishing distinct sport tour destinations. French resorts, in particular, have been developed for the mass market, and as Mintel (2002) points out, they have mostly been purpose-built and designed to provide a high level of convenience for the skier. However, convenience has come at the expense of atmosphere, and French resorts lack the Alpine charm of more traditional resorts in Austria and Switzerland. French resorts also provide a broad range of accommodation, and while much of it is provided in apartments instead of hotels, and the apartment blocks, built during in the early development of the mass-market ski industry, are cramped and fall short of today's tourists' expectations-a feature that is similar of much apartment development in many coastal Mediterranean mass tourist resorts.

A complementary facility has also been shown to be important for 'activity tourism,' which includes a variety of outdoor pursuits undertaken for physical pleasure. There are various types of facilities and accommodations available, ranging from specialized centers run by local governments, schools, and voluntary organizations, and also the private sector, to small private concerns including guest houses, farm-based accommodation, and caravan and camping sites. The specialist activity center, which, while specializing in specific activities, also offers opportunities for a much broader range of pursuits, is a distinguishing feature of activity tourism. According to Clark et al. (1994), such multi-activity centers offer a diverse range of activities, with the four most common being rock climbing, orienteering, skiing, and canoeing. Plas-Y-Brenin in North Wales, for example, has a training wall, an artificial ski slope, and a canoe pool, and offer training in rock climbing, summer mountaineering and hill walking, winter climbing and mountaineering, alpine skiing and mountaineering, mountaineering qualifications and awards, kayaking, and canoeing, and paddling. In addition to coursework, it provides lodging for those who wants to travel the area and make their own activity arrangements. Multiple sport activity vacation centers, which have grown in popularity in recent years, are not limited to upland areas.

2.5.3. Political and Economic Resources

As previously stated, the built environment is an important component of the sport tourism destination, even in so-called "resource-based" environments. The development of both sports facilities and related infrastructure necessitates capital investment; thus, the quality of destinations will be determined by the extent to which the private and/or public sectors are willing to invest in such locations. Historically, the state has been more willing to invest in sports facilities while being less willing to invest in tourism. However, given the importance of both sport and tourism in modern societies, this position has shifted, and governments have recently been equally interested in developing tourism-often in collaboration with private enterprise- and sport tourism has been especially important. The city has been one of the primary destinations for such investment. Cities offer a variety of incentives for the growth of sport tourism, particularly in relation to spectator sports. Given their size and market influence, they invariably have high-quality facilities and stadia, which were originally built for their own residents and sports teams. Domestic and subsequent international competitions, on the other hand, have encouraged a large number of people to travel to cities, either to support their teams or to enjoy the sporting spectacle. Furthermore, cities are increasingly hosting mega-sporting events, both using existing facilities and building new ones specifically for this purpose.

According to Weed and Bull (2004), larger cities are particularly prominent, with their sports teams attracting greater potential support and subsequent income, which in turn attracts financial investment from both local and external businesses, thus helping to maintain future success and support. Because of their large local catchments, transportation hubs, and general tourist infrastructure, large cities, particularly capital cities, are logical destinations for national stadia. Various sports have national stadia in London, with obvious examples including Wembley, Twickenham, Lords, and Wimbledon, though other UK cities are also able to host national and international events (e.g., Meadowbank in Edinburgh, the Millennium Stadium in Cardiff, and the Don Valley Stadium in Sheffield). Cities, on the other hand, have recently become even more important sport tourism destinations as a result of two overlapping influences. The first is an acknowledgement of the importance of mega-events (including major sporting events) in shaping both the national tourism product and the long-term city product, and the second is how sport in general has been used as part of tourism's role in urban regeneration.

Special sporting events have emerged as important tourism policy tools for governments seeking to boost local business through visitor spending. They not only generate significant revenue, but they are also viewed as a means of improving the image of both the city and the state as a whole.

2.5.4. Cultural and Perceptual Aspects

Sport, as many writers have pointed out, is a cultural form, as is tourism. Sport, according to Standeven and De Knop (1999), is a cultural experience of physical activity, whereas tourism is a cultural experience of place. However, in addition to defining these two distinct cultural experiences as separate spheres, it must also be the case that the two are associated when the resources that people believe they require are not readily available in their immediate surroundings. This blending of different cultural experiences complicates the assessment of resource quality because it is linked to how people evaluate and perceive resources in general. It could be argued that all resources are cultural appraisals in some way, and nowhere is this truer than in various sporting environments. While many people can easily pursue their sport near their home, they frequently choose to travel elsewhere to participate in what may be considered a preferred environment. This is not because the standard of the facility is necessarily higher elsewhere, but rather because of the atmosphere of the place-the climate may be better, the environment less polluted, less crowded, or more peaceful, or the general landscape more scenically appealing-but for some spectators, being part of a raucous crowd may also be part of the attraction. Urry (1992) adds another point about performing familiar activities in unusual visual environments. Swimming and other sports, he says, "all have special significance if they take place against a distinctive visual backdrop." The visual gaze transforms ordinary activities into extraordinary ones.' The illustration of cyclists choosing to cycle through rural instead of urban areas has already been mentioned, but another example is British golfers who may prefer to play golf in Southern Spain or the Algarve rather than, or in addition to, using courses at home.

The perception and experience of such environments by sport tourists is clearly contentious. Playing sport in environments that improve performance will undoubtedly be appealing to some, particularly elite athletes, whereas for others, pitting their skills against the elements may be part of the appeal. Similarly, for spectators, 'safer,' immediately identifiable environments may be preferable, and just as people travel to expertise spectacle in other spheres, this may be the primary attraction for many sport tourists. Nonetheless, some have argued that the blandness of stadiums, the more standardized crowd experience, and the greater distance spectators become from the action of the sport in ever-larger stadiums may make it less appealing to some.

2.6. POLICY AND PLANNING FOR SPORT TOURISM DESTINATIONS

Sport tourism planning is the establishment of sporting facilities, events, and critical facilities to attract visitors and users from outside the immediate region (destination) as well as local users. However, when considering sport tourism planning, consideration is given not only to physical considerations of planning, but also to social, environmental, and economic goals, as well as the long-term implications of sport tourism development.

The achievement of socio-cultural, economic, and environmental goals is required for the long-term development of sport tourism. An effective sport tourism economy should preferably support and enhance the social/ cultural dimension of the destination, as well as the natural and built environment, which plays an important role in many types of sport tourism. The interaction between sport and tourism, on the other hand, is not always positive. Numerous examples of sport tourism show the high potential for negative effects on host communities. As a result, the development of sport tourism development at tourism destinations necessitates the use of constructive and integrated policy and planning approaches. Where such approaches exist now, they are the exception rather than the rule.

Planning is a method of managing change by making decisions that will guide future actions. In the context of sport tourism, planning is defined as "the process of anticipating and regulating change in a system in order to promote orderly development and increase the social, economic, and environmental benefits of the development process." The underlying planning process is based on an assessment of the current situation, anticipated changes in the environment in which the plan is being implemented, decisions on the desired end state, the formulation of some sort of action plan, and its implementation, followed by monitoring, assessment, and adjustment as needed (Figure 2.3).

Figure 2.3. Sport tourism has gained widespread acceptance as a means of generating economic growth through inward investment, tourism, and the development of service sector interests.

Source: https://webunwto.s3.eu-west-1.amazonaws.com/styles/article_main_image/s3/2019-09/sport-tourism-big.jpg?itok=tlWBN2sm.

Despite the potential of sport tourism, systematic or strategic planning in sport tourism is uncommon. Sport tourism developments are often pursued in the interest of short-term economic returns, with long-term strategic visions being less clear. This is unfavorable because many sport tourism events are "short-term events with long-term consequences for the cities that host them." In many cases, the long-term consequences for sports mega-event destinations are negative. The potential for sporting events to have negative consequences grows in proportion to their size. Thus, in the absence of careful planning, the long-term legacies of sporting events may be a financial burden borne by local ratepayers.

2.6.1. Need of Planning for Sport Tourism

The emphasis on short-term economic outcomes, at the expense of long-term evaluations of sport tourism and its effects on destinations, is the result of a number of factors that challenge the sport tourism planning process. Planning takes place at several levels, along with strategic and operational. Sport tourism planning also occurs in a variety of domains, such as facility development, infrastructure, event planning, and marketing. Each planning domain may be engineered by various public or private organizations, and the stakeholders who may influence (or be influenced by) these planning processes vary. Furthermore, sport, and tourism planning occur at a variety of spatial levels, ranging from international to local site plans. Thus, sport tourism planning may take place in distinct policy communities, sectors, and at the international, national, and regional/local levels. Sport tourism plans require coordination and support from a wide range of public and private sector stakeholders, and they must be involved in the planning process. The likelihood of success is increased, as in any planning process, when key stakeholders have a clear understanding of the potential that sport tourism offers destinations. As debates in sport tourism become more advanced, such visions become more focused.

Because of the widespread interest in sports, major national and international sporting events draw tens of thousands of spectators and visitors. The often-extensive media broadcasting may increase the popularity of sports and provide tremendous opportunities for the marketing and imaging of destination areas. Mega sports events, such as the Olympic Games, the Super Bowl, and various World Cups of football, rugby, and cricket, for example, have considerable implications on host cities and regions before, during, and after such events. The sporting legacy of new facility development can have a long-term impact on communities, visitors, and athletes.

The tourism benefits of a well-developed facility or a well-staged event could be extremely valuable for the potential increase in the number of inbound tourists. In the case of Winter Olympics hosts, for example, the venue and hosting towns often get highly identified in the winter tourist marketplace and can achieve a high level of international awareness of their attractiveness in winter-a time that is often low season for tourism. For example, Norway's hosting of the Olympic Games in 1994 was viewed as an opportunity to promote activities in the country's mountainous regions, such as skiing, ski touring, and mountain climbing, which could attract more domestic and international tourists.

Government and industry have almost universally sought the perceived potential of sport tourism and sports events to contribute to economic growth and attract investment and tourism. However, in the rush to attract tourists, 'relatively little thought is usually given to the nature of the planning process with which to maximize the potential of tourism for the host community.'

Sport tourism planning takes many forms (development, infrastructure, labor force, promotion, and marketing), many public structures (different government and industry organizations), and many scales (international, national, regional, local, and sectoral). The emergence of public concern about the social, economic, and environmental externalities associated with tourism development has resulted in calls for government regulation and planning of touristic activity, including sport tourism. As a result, sport tourism planning should be concerned with anticipating and regulating the effects of the development of sport tourism-related events, facilities, and infrastructure on the host community, as well as promoting associated development in a way that maximizes short and long-term economic, environmental, economic, and social benefits. However, it must be stated right away that long-term assessment of sport tourism and its effect on destinations is virtually non-existent. It may surprise many readers who think that planning should be a rational process which includes such evaluations, especially given the future orientation of the planning function. However, the 'messy' nature of sport tourism-related planning has long been recognized. One of the great myths about such events, according to Armstrong, who examined 30 international mega-events and projects, many of which included sport, is that they are planned in a rational manner.

The project planners' main responsibility was to provide support for previously made decisions. The following traditional stages are excluded from the prestige project planning process: recognition and definition of the problem, definition of the planning task, data collection, analysis, and forecasting, constraints determination, testing of alternatives, plan evaluation, and project evaluation.

Sport tourism planning, particularly for large sports events and facilities, has generally followed a top-down style of planning and promotion, leaving tourism destination communities with little input or control over the community's future. Arguably, the larger the sport tourism event or facility, the less input local users have, and it appears that only local leisure and recreational facilities receive some consultation. The results of this central planning style have been mixed. Tourism has contributed to rapid economic growth and regional development in some cases, but its accompanying impacts and spillover effects are frequently overlooked. Failure of government, industry, and planners to meet community needs and concerns can result in negative resident attitudes toward tourists and the tourist industry, as well as tourist dissatisfaction with the destination. However, if the government and private sector want to use sport tourism as a mechanism to attract tourism and investment, such negative attitudes simply cannot be tolerated in the long run. This is due to the fact that tourism, more than any other industry, is dependent on the goodwill and cooperation of local people because they are a component of the product. Where development and planning do align with local aspirations and capacities, resistance, and hostility can raise the cost of doing business or completely destroy the industry's potential.

As a result, careful analysis of the dynamics of sport tourism planning is a sufficient prerequisite for the successful achievement of destination-level sport tourism development objectives. In an ideal world, destinations would use sport tourism as a means of development not for its own sake, but because it is a means to an end in terms of returns to the destination. Planning is a critical tool for achieving such goals.

2.6.2. How to Create an Effective Destination Planning Process?

2.6.2.1. Key Elements in Destination Planning

A pragmatic approach to destination management is a multi-level strategy that highlights a destination's success from multiple angles. This entails various levels of tourism planning, such as destination marketing and tourism management, as well as having a clear vision for future tourism development. Here are the five most important steps in the tourism planning process:

- Determining the form and size of tourism, as well as how it will emerge for the destination;
- Researching the most effective and sustainable direction for the destination's future development;
- A broad-scope, strategic approach to tourism marketing;

- Creating critical planning documentation;
- Working with key stakeholders to collaborate, agree, and engage.

2.6.2.2. Involving Stakeholders in the Destination Planning Process

Stakeholders play an important role in the destination planning process. Most stakeholders want a tourism management plan based on solid fundamental principles that have been thoroughly researched and are based on solid analytics.

It is also critical to convey to stakeholders a strong sense of leadership. Stakeholders expect a well-defined planning and management process based on reliable information. Finally, stakeholders want to get to know the team and know that the tourism management plan has a vision for the destination's future as well as the leadership to make it a reality. This is central to the stakeholder engagement process and is necessary for developing a shared vision and enhancing communication.

2.6.2.3. Strategic Planning Documents in the Destination Planning Process

Strategic planning documents developed during the destination planning process should lay the groundwork for the destination's tourism growth and management. For tourism management, this entails using a broad lens to plan for the destination's long-term direction. Documents for strategic planning should include:

- A well-defined long-term vision for tourism management that clearly specifies priorities.
- A well-researched plan identifying the data and analytics required to engage stakeholders in making informed decisions.
- Risk assessments that identify the threats and hazards that may affect the destination's tourism industry, as well as how the destination will respond in the event of a crisis.

2.7. POLICY COMMUNITIES FOR SPORT AND TOURISM

Sport and tourism policy development can be thought of as taking place in sectorized arenas known as policy communities. A policy community refers to groups that share an interest in a policy sector and interact with one another on a regular basis in order to reap the effectiveness of their relationships. Government departments and agencies, commercial sector firms, voluntary clubs and associations, charities, and any other person or group with an interest in the development of policy in a specific sector are all included.

Sport and tourism policy communities exist within a larger leisure policy universe that includes a large number of organizations, groups, associations, and individuals who share a common interest in leisure policy and may attempt to contribute to policy development on a regular or irregular basis. A simple Venn diagram (Figure 2.4) depicts the location of sport and a network to deal with sport-tourism issues.

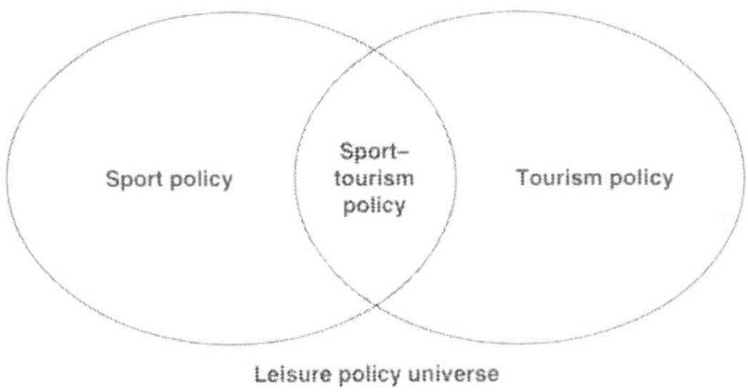

Figure 2.4. Sport and tourism policy interactions.

Source: http://www.microlinkcolleges.net/elib/files/undergraduate/Tourism%20&%20Hotel%20Management/Sports%20Tourism%20Destinations%20(2005).pdf.

A policy network is defined by Wright (1988, p. 606) as "a linking process, the result of those exchanges within a policy community or between a number of policy communities." Furthermore, Wilks and Wright (1987) hypothesize that a major advantage of the policy network concept is that it enables members to come from various policy communities. This is especially useful in investigating the type of cross-sectoral policy liaison required for collaborative sport-tourism policy development.

Another advantage of the model depicted in Figure 2.4 is that it allows for an examination of the impact of structural issues within individual policy communities for sport and tourism on the emergence and development of collaborative sport-tourism policy. These examples highlight the more favorable conditions for the development of collaborative policy, first at the subnational region or small-state level, and then at the city level.

2.7.1. The Operation of Policy Communities for Sport and Tourism

In the structure and operation of policy communities, there are a number of common salient issues. These are about leadership, membership, resource reliance, and structural relationships. Leadership is a thorny issue in many sports and tourism policy communities around the world. The Department for Culture, Media, and Sport would be expected to be the lead government department in the UK; however, historically, policy issues for sport and tourism have been devolved to partially autonomous, 'arms-length government agencies, such as the Sports Council and Visit Britain. The Department for Culture, Media, and Sport formed a Tourism Advisory Forum in 1997, comprised of prominent figures from the tourism industry, to advise it on tourism matters, which is precisely the role the English body was established to fulfill in 1969. The formation of the Tourism Forum was a clear first step toward the replacement of the old English Tourist Board with the English Tourism Council in 1999, and the subsequent subsumption of the English body into the British organization in 2003. The Department has also increasingly limited the Sports Council's autonomy, to the point where its protracted restructure into United Kingdom and English Sports Councils has contributed to the new Sport England spending much of its time depending on distributing Lottery Sports Fund grants in accordance with government departmental principles. Sport England's direction is now much more under the control of the Department for Culture, Media, and Sport, and as a result, the organization has become an agent instead of an instigator of national sports policy. This condition brings conflicts between governments and national agencies in both the sport and tourism policy communities.

2.7.2. Strategic Planning of Sport Tourism Events on Destination Brands

Over the years, research on destination branding has focused on the role of branding in the consumption of tourism destination brands. According to such studies, the sport events industry is one of the most important industries that serves as a catalyst for branding opportunities in which destinations can reap the benefits of tourism. However, the globalization of sport, along with the globalization of mass media, has increased marketing demands on tourism destinations to not only promote their brands through an organization with sport but has also accelerated competition among destinations for a share of the domestic and international tourism markets that facilitate sporting events.

Because the concept of destination branding arose from general branding literature, it has frequently been viewed as running parallel to corporate, product, and service branding theories (Balakrishnan, 2009). However, while destination branding is commonly understood on the basis of generic branding, it frequently only refers to the modern form of tourism promotion (Anholt, 2005; Hanna and Rowley, 2008; Boisen et al., 2018). A widely accepted definition that constructs on the premise of the generic branding theory to promote travel and tourism, describing destination branding in terms of its unique and identifiable characteristics in its name, logo, and any other word or graphic; as well as the pledge of delivering a noteworthy travel experience that is mainly linked to unique associations of the destination (Ritchie and Ritchie, 1998). According to more recent studies, destinations not only create memorable travel experiences for the benefit of unique associations, but they also seek to position their brands in such a way that visitors' experiences of a destination will result in repeat visits as well as voluntary ambassadors for the destination (Morgan et al., 2010). As a result, destinations are increasingly viewing branding as a more strategic tool for brand positioning (Filo et al., 2013). Furthermore, Greaves and Skinner (2010) contend that developing a distinct brand identity for tourism destinations will assist current and potential tourists in differentiating one destination from another. In line with this statement, Kapferer (2012) contends that it is critical for a place brand to have a clear value proposition that serves as the brand identity, in order to facilitate the derivation of the associated benefits. Sport (events) are widely regarded as powerful value propositions for positioning brand identities and achieving various branding benefits (Govers and Go, 2009; Zhang and Zhao, 2009; Hemmonsbey and Knott, 2016). Melbourne, Australia, is a prime example of a city that has positioned itself as one of the world's leading destination brands as a result of its brand identity (logo), which features sport as one of its portfolios. Hemmonsbey and Knott (2016) discovered that Cape Town, South Africa, has recently redeveloped its identity (in the form of its logo and slogan), with [sport] events being one of its investment and marketing strategies. In the face of globalization, tourism destinations should develop a niche market and clearly distinguish themselves from other competing destinations, according to Morgan et al. (2010). Cities, towns, villages, and nations are examples of such destinations (Camilleri, 2019). It is clear from the preceding discussion that the broader context of destination branding extends to brand differentiation applications in order to achieve not only brand awareness but also brand positioning. Furthermore, in addition to establishing a clear brand identity, destinations should consider including a value proposition in the form of a special niche area that will benefit and sustain the tourism destination brand. Sport events will be investigated here as a tourism niche that is strategically planned and capitalized in order to derive tourism and destination branding outcomes.

2.7.3. Sport Events as a Means for Strategic Destination Branding

The nature of sporting events usually provides opportunities for host cities to engage in branding practices, allowing cities to increase brand awareness and communicate positive messages about their city to the rest of the world. Sport event hosting also provides opportunities for economic and social development. To achieve long-term destination branding benefits, additional opportunities for strategic planning through stakeholder coordination of event leveraging are implied. Sport mega-events, such as the Olympic Games and the FIFA World Cup, have found to be an excellent feature in positioning nations to the international, or global, society due to their large-scale and international hosting nature (Horne and Manzenreiter, 2006). Mega-events have also been shown to have significant legacy implications for the host and non-hosting communities, which can be negative or positive, planned, or unplanned, tangible or intangible (Swart and Bob, 2012; Nauright, 2013; Kim et al., 2015; Lui, 2016). Such legacies have an impact on sports, cities, infrastructure, the economy, the environment, politics, and society (Cornelissen et al., 2011). For example, the 2004 Olympic Games provided unique opportunities for Athens and Greece to diversify and enhance their destination's tourism product, as well as their successful Olympic Games. Organization has left important tangible and intangible legacies for Athens linked to its heritage and tourism infrastructure (Ziakas and Boukas, 2014). Aside from that, the 2008 Olympic Games left a number of legacies in Beijing and throughout China. Social legacies included public health, education, and volunteerism, as well as urban and environmental legacies such as road and transportation infrastructure development, environmental sustainability, cultural/heritage preservation, and venue/facility upgrades (International Olympics Committee, 2009). Another example is the 2010 FIFA World Cup in South Africa, which adopted an African Legacy Program in order to maximize the World Cup's impact on the African continent as a whole (Tichaawa and Bob, 2015). While the effects of sport on destinations are visible in the legacy effects following event staging, Tichaawa and Bob (2015) argue that authorities must plan ahead for long-term impacts rather than assuming or anticipating such impacts to materialize. In the case of Athens, Ziakas, and Boukas (2014) observe that, despite the success of the Olympic Games' organization, there was a clear lack of coordination between commercial providers and public governing bodies in terms of a joint strategic approach to leveraging the Games. As a result, the dream of utilizing post-Olympic facilities was dashed. Chalip (2017) presents a forward-thinking theory, arguing that the quality of impacts derived from sport mega-event hosting is dependent on a coordinated approach to event leveraging. This advanced approach can either be a new challenge for both host and non-host countries or, if properly implemented, can maximize the potential for positive, long-term benefits (legacies) while minimizing the risks to tourism and destination branding. Regardless of the above-mentioned ideas about the impact of mega-events on destination branding, it is worth noting that there are a limited number of mega-events, which leads to strong competition among destinations in auction to host such events. As a result, the extent to which the full extent of the impacts is realized is limited to a few hosting countries. As a result, the home-grown event is being introduced to provide a diverse range of destinations with an equal opportunity for a share of the global tourism market and, as a result, to derive the intended brand-related benefits. Home-grown events, according to Hemmonsbey and Tichaawa (2018), are recurring events that are smaller in size and scope than mega-events but have a strong association with the destination's brand identity. Such events are distinguished further by their widespread (local and international) participation and their inherent characteristics in terms of their actual staging, which takes place in an outdoor, natural setting. It is assumed that home-grown events are conceptualized, produced, and nurtured by a specific location. Destinations, according to Brown, Chalip, Jago, and Mules (2004), strategically use events of this type to strengthen their existing brand image because the event's associations are consistent with one or more aspects of the destination. As a result, destinations benefit from events that highlight their beaches, landscape, or climate, such as Cycle Tours, water sports, triathlons, and mountain bike races, amongst others. It is obvious that events come in a variety of shapes and sizes, and stakeholders may include one or more types of events in their strategic process in order to achieve tourism destination benefits. Indeed, Chalip (2004); Ziakas and Costa (2011); and Ziakas (2018) encourage destinations to include a multi-event event portfolio in their strategic initiatives. This study contextualizes the home-grown event for destinations as part of their strategic planning and overall leveraging practices, in light of this viewpoint and in relation to the home-grown event theme.

2.7.4. Strategic Planning Through Sport Events Leveraging

Strategic planning, in essence, serves as the foundation for leveraging in its approach to recognize events as opportunities for designing and implementing strategies to achieve desired outcomes. Earlier studies on the strategic planning of sporting events argue for the importance of evaluating planned strategies prior to and after event hosting. This viewpoint is supported by advanced research on leveraging, which refers to leveraging as a strategic ex-ante, analytical approach as opposed to an expose, impact-driven, outcomes approach (O'Brien and Chalip, 2008; Smith, 2014). As a result, both planning techniques aim to produce a 'forward-thinking... strategic approach,' in which both the impacts and the methods for achieving them are planned ahead of time. Furthermore, strategic planning through leveraging advocates for the need to develop strategic objectives that contribute to a larger goal. Given a clear set of strategic objectives as a result of strategic planning, stakeholders are said to be able to take ownership of their chosen objectives and approach. However, there is still a need to define the roles and responsibilities of stakeholders in stakeholder group collaboration for optimal leveraging. Regardless of circumstance of stakeholder collaboration, the case study of sport event leveraging for Australia and the 2000 Olympic Games provides best practices for stakeholder collaboration prior to and after event hosting. In an effort to strategically position Australia as a leading destination for sport tourism, stakeholders from both public and private institutions combined their efforts and goals in pursuit of a larger goal. Stakeholders' leveraging efforts have also enabled this country to capitalize on Olympic media coverage of their brand image, which has resulted in stakeholders influencing respective journalist stories about their destination brand. Grix (2012) reported on Germany's attempts to engage in sport event leveraging to improve their nation's (poor) image abroad in a different case where destination brand image was the driving force for strategic planning. This discovery resulted from international perceptions of the people as "dominant," "arrogant," and "dull," as well as residual Second World War and Nazi perceptions, particularly in the British press (Grix and Lacroix, 2006). Thus, these two examples contradict the earlier mention of Athens and their failed attempts to coordinate stakeholder efforts toward a joint strategic approach to leveraging the 2004 Olympic Games. The discussion on strategic planning through sport event leveraging emphasizes the deliberate design and implementation of marketing strategies, the formulation of strategic objectives, and the formation of stakeholder partnerships. Case studies show that engaging with global media networks in relation to the image, product, and service offering of the host destination coherently positions tourism destination brands in such a way that long-term positive perceptions are formed. As a result, these strategic planning approaches are investigated in the context of developing destinations through the hosting of annual, recurring events.

2.8. TOURISM DESTINATION MANAGEMENT

Tourism destination management is an important part of reducing tourism's environmental impact. Land use planning, business licenses and zoning controls, environmental as well as other regulations, business association policies, and a variety of other techniques can all be used to shape the development and day-to-day operation of tourism-related activities. The idea that a single, recognizable destination may also include several municipalities, provinces, or other government entities-in island environments, the entire country-complicates destination management.

"Destination management organizations" (DMOs) are participatory governance structures led by local authorities, with the participation of local NGOs, community, and indigenous representatives, academia, and local chambers of commerce. Local tourism boards, councils, or development organizations are frequently used as DMOs. A destination's network of local tourism businesses (hotels, attractions, transportation services, service providers such as guides and equipment rentals, restaurants, and so on) is also an important component. Tourism's needs, expectations, and anticipated benefits vary greatly from destination to destination, and there is no "one size fits all" approach to destination management.

Chapter 3

Destination Marketing and Sport Events Management

CONTENTS

3.1. INTRODUCTION

Tourism is one of the world's fastest growing industries. Sports tourism is one of the fastest growing forms of alternative tourism in the tourism industry. Tourism destinations have been focusing on tourism diversification in order to comply with sustainable tourism policies. Sports tourism, in this context, is a form of sustainable tourism that improves the health of local people, has positive social effects, and boosts the marketing power of destinations. Locals will benefit from the infrastructure developed for sports tourism, and the society's sports culture will grow as a result. The involvement of locals and visitors in joint sporting events would ensure that the social benefits of tourism are maximized. If these sporting events reach a certain level of quality and ensure the satisfaction of the participants, and the participants' loyalty to the events can grow over time. Sporting activities can also help a destination's branding. For destination marketing, branding is a powerful tool. The Olympic Games, the world's largest sporting event, provide significant brand value to the cities in which they are held. Tourism-related sporting activities are an appealing component for destinations, and when well-planned, they offer strategic benefits in destination management.

3.2. DESTINATION MARKETING

Destination marketing can be extremely beneficial to anyone who stands to profit from increasing the number of people who visit a particular venue.

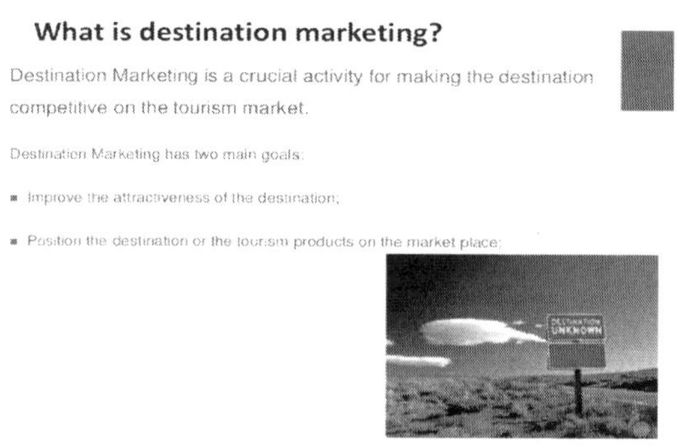

Figure 3.1. What is destination marketing?

Source: https://th.bing.com/th/id/OIP.t0_Wtbd10RYrmgcw4ZNJVQHaFj?pid=ImgDet&rs=1.

This involves local and national governments, airlines, travel agencies, as well as entertainment venues, tourist attractions, and hotel and resort operators. You can learn more about some of the destination marketing techniques that can help to draw more tourists in this section (Figure 3.1).

3.2.1. What Exactly Is Destination Marketing?

Destination marketing is a form of marketing that promotes a destination to potential tourists in order to increase the number of people who visit that venue. The destination may be very specific, as in town or city marketing, or much wider, as in area or country marketing.

Finally, the goal of destination marketing is to make your place or destination seem more appealing than the key alternatives, increasing the number of tourists who visit and assisting the local tourism industry. It is also about raising the visibility of the destination in order to increase demand and improve its credibility.

3.2.2. What Is the Importance of Destination Marketing?

It is important to emphasize the benefits of a particular destination in order to entice travelers to visit it. This is particularly important as one destination is vying for the same tourists or visitors as another. Both the Caribbean and the Great Barrier Reef, for example, have excellent diving opportunities (Figure 3.2).

Figure 3.2. The aim of most destination marketing is to endorse destination as superior to alternatives by emphasizing the items that distinguish it or make it a desirable place to visit.

Source: https://image.slidesharecdn.com/destinationmarketingyac-110520111020-phpapp02/95/destination-marketing-yac-9–728.jpg?cb=1305890066.

Effective destination marketing will then be extremely beneficial to the local tourism industry, which includes hotels, restaurants, bars, airlines, and other related suppliers.

3.2.3. Who Is Responsible for Destination Marketing?

Destination marketing is usually handled by a dedicated destination marketing organization (DMO) or tourist board. These organizations serve the group or destination being marketed and use a variety of marketing tactics to pique the interest of travelers in the area (Figure 3.3).

Figure 3.3. In several ways, a DMO acts as a connection between the visitor and the destination, not only promoting the visit but also providing valuable visitor information.

Source: https://www.iteratemarketing.com/wp-content/uploads/2016/01/Destination-Marketing-Organizations-Are-They-A-Waste-FB.png.

Hotels, local companies, attractions, and their owners are often members of a DMO, and support is typically provided by the government and/or membership fees.

3.2.4. Destination Marketing Strategies to Attract More Visitors

A variety of marketing techniques can be used for destination marketing, assisting in raising overall awareness of a destination and the number of people who visit. More information about these techniques, as well as how destination marketing organizations should use them, can be found below:

1. ***Identify Your Unique Selling Points (USPs):*** Taking the time to clearly identify what makes your destination special is arguably the single most important move for any organization engaged in destination marketing. There are many ways for a destination to stand out, such as offering unique experiences to tourists or having unique natural features, such as mountains, beaches, or volcanoes.

   It is possible that your place has an interesting past or notable landmarks, such as the Eiffel Tower, the Empire State Building, or the Great Wall of China. Perhaps the location's culture is its unique selling point, and tourists will want to travel to see local festivals, museums, or sports clubs. Make an effort to find as many USPs as possible.

2. ***Identify Your Target Audience and Market:*** Another effective destination marketing strategy is to recognize the target markets. Consider who is going to want to visit the destination and why. Sometimes, you will need to segment this into many different groups, each of whom will have a different reason for wanting to visit your place.

   For example, it is possible that your destination appeals to youngsters on a gap year and elderly couples enjoying their retirement. It can cater to individuals who love outdoor sporting activities as well as those who want to relax on the beach. You must also remember the various classes, such as visitors, business travelers, and students.

   Finally, consider why some markets are more likely than others to be involved. This may include people from neighboring countries if you have good transportation, or people who speak the same language.

3. ***Make Use of Data for Analytics:*** Destination marketing organizations have the ability to collect and use massive quantities of data for a variety of purposes. For example, the organization's website could allow you to use tools such as Google Analytics to learn more about your visitors, such as who they are, where they came from, and what motivated them.

Offline, you can learn about current travelers, such as their average age, whether your destination prefer men or women, and what modes of transportation they prefer. If you have collected enough information, you can analyze it to determine the best people to contact, the best ways to contact them, and the best messages to send.

4. ***Promote Your Destination:*** Branding is a tactic used by firms to make themselves more readily recognizable. Logos, color schemes, and other design concepts, as well as slogans or the frequent use of specific terminology, may all be considered branding. Finally, branding is about being distinguishable and standing out from the crowd.

   This theory is fairly simple to apply to a destination. Create a color scheme that is consistent, use a tagline that says something about the destination, create hashtags for people to use on social media, and be as consistent as possible with the advertising messages so that people become acquainted with it.

5. ***Involve All Stakeholders:*** An important aspect of destination management is looking out for the interests of different stakeholders in order to build confidence and promote interaction. Officials from your country, city, village, or state, as well as hotels, attractions, restaurants, shopping malls, travel agents, and tour operators, may be among the stakeholders.

   Develop destination marketing strategies that allow these different parties to engage in and endorse your destination. See if you can persuade stakeholders to use some of your overall destination branding and to run their own advertising or marketing campaigns in order to increase visitor interest.

6. ***Design a Fantastic Destination Website:*** A destination website is an excellent place to highlight your destination's USPs, communicate directly with your target audience(s), and promote your destination through images, videos, and virtual reality tours.

   You can even use the website to provide the travelers with all of the information they require ahead of their trip, as well as to promote lodging options, attractions, events, places to eat and drink, and other services. It is critical that your website is mobile-friendly, and an on-site blog will help to keep visitors coming back.

7. ***Search Engine Optimization:*** When people utilize search engines like Google to search for reasons related to your destination or features that your destination can provide, you want your website to appear near the top of the search engine results pages. The most effective way to accomplish this is to develop a comprehensive search engine optimization strategy (Figure 3.4).

Figure 3.4. SEO entails researching keywords, creating content that targets those keywords, and employing a variety of other techniques to boost your ranking.

Source: https://th.bing.com/th/id/OIP._i6LdMq_3l-Gg58-DFpYwgHaGL?pid=ImgDet&rs=1.

SEO principles can also be used to promote videos and images. An on-site blog can also help here, as it will provide you with plenty of fresh content to optimize with strategic keyword usage.

8. ***Experience Marketing:*** Most travelers are motivated by their experiences, so marketing a destination in this manner makes sense. Instead of displaying attractions, use promotional material to demonstrate how people interact with them. One method is to encourage the sharing of user-generated content so that people can share their own moments on your website and social media.

When you accomplish this, your own visitors become ambassadors for your destination. Word-of-mouth marketing can be extremely effective because it is perceived as more reliable than hearing from an organization with an obvious purpose.

9. ***Video and Virtual Reality Marketing:*** Both video marketing and virtual reality marketing will go a long way towards improving your destination marketing efforts. Video content can be easily shared through networks including Facebook, Twitter, Instagram, and YouTube. These videos will promote local points of interest or feature locals discussing your destination (Figure 3.5).

Figure 3.5. Virtual reality marketing.

Source: https://th.bing.com/th/id/OIP.JGdEy32vyzqsS7eG4-nrYwHaD4?pid=ImgDet&rs=1.

Virtual reality marketing takes it a step further by allowing your target audience to explore facets of your destination from the comfort of their own home. A virtual tour of a nearby hotel, a virtual travel experience of a nearby entertainment venue, or 360° tours of attractions or landmarks could all be examples.

10. ***Social Media Marketing Techniques:*** Social networking sites such as Facebook, Twitter, and Instagram are excellent ways to reach out to people and promote your destination. In fact, the destination marketing strategies available here are nearly limitless, ranging from simple promotional posts, photos, and video content to competitions and even viral content or memes.

Furthermore, the majority of social media sites have paid marketing opportunities, such as ads or supported content. These will increase the reach of your social media marketing campaigns and can be targeted at very specific audiences, allowing you to target people based on age, gender, place, and even online browsing habits (Figure 3.6).

Figure 3.6. Social media marketing.

Source: https://th.bing.com/th/id/OIP.Fht2ISTe4GE1aBu_v0HZEgHaFK?pid=ImgDet&rs=1.

11. ***Collaborate with Influencers:*** Many modern internet users have a natural aversion to blatant marketing tactics such as direct ads. Many of these individuals, however, can be reached indirectly through influencers. Influencers are people who have a large audience on sites such as Instagram, Twitter, YouTube, Facebook, or their own blog.

Their followers are loyal and trust the influencer's views more than they trust an organization. By collaborating with influencers, you can promote your destination through multiple platforms, including to niche audiences, who are far more likely to react positively to the message being promoted.

12. ***Advertise Your Destination on Travel Websites:*** In addition to getting your own destination website, it is critical to be able to meet people who are unfamiliar with your destination. International travel websites, such as TripAdvisor, as well as local websites or websites targeted at particular audiences, offer a great answer.

    Travel websites can be used to advertise your destination in a variety of ways, including sharing photos, publishing articles, producing video content, and even promoting virtual reality tours. Travel websites can also be used to track ratings, react to criticism, and enhance feedback over time.

13. ***Online Advertising Strategies:*** You can use the knowledge and data you have learned about your visitors and target audience to guide your online ads. Using this data, you can tailor search ads to specific areas, pay for display advertising on the appropriate platforms, and promote content on third-party websites visited by your target audience.

    Furthermore, re-marketing enables you to contact people who have already interacted with your destination website or your DMO's social media networks. This allows you to remind people about your destination while recognizing that they have already shown interest, which is perfect for motivating them to commit to a visit.

14. ***Offline Promotional Techniques:*** You should be able to predict where your target group is going to move until you understand who they are, what they want, and where they are coming from. Some destinations, for example, draw visitors during the summer months, while others draw visitors for winter sports. Some could be more appealing during the holidays, such as Christmas.

    This then makes developing a detailed offline advertising plan simpler. Offline strategies may include television ads, radio advertising, and promotion in newspapers and magazines. You could pay for advertising, write stories about your destination, and target specific local publications in areas where your target audience lives in the latter two examples.

3.3. MEANING AND DEFINITIONS OF EVENTS

Events are a competitive and rapidly growing industry with clear synergies with tourism. They can expand the visitor economy, provide media attention, encourage regional growth, and stimulate infrastructure upgrades and the formation of new alliances for financing sport, tourism, culture, and leisure facilities if they are handled and hosted effectively. Major event hosting thus provides a once-in-a-lifetime opportunity to rethink or reposition a destination while still supporting the growth of modern infrastructure. As a result, many countries now see efficient hosting of such events as a means of generating economic growth, job creation, branding, well-being, and urban regeneration. Many cities, regions, and countries are now committing significant resources to creating, attracting, and promoting major events as part of a larger strategy to increase tourist numbers and spending (Figure 3.7).

Figure 3.7. Events enable people to interact with their surroundings, spend time together, celebrate, and experience the diversity of cultures, and promote creativity and innovation.

Source: https://th.bing.com/th/id/OIP.gVI1lez6xw0upjCTTIKLwAHaE8?pid=ImgDet&rs=1.

They bring a community to life and encourage a destination to highlight its tourism experience while increasing economic activity. Community building, lifestyle, and recreation enhancement, cultural advancement, tourism promotion and increased visitation, volunteer engagement, fundraising, and economic development are all aided by events. Above all, events foster a sense of fun and vibrancy, resulting in a clear sense of group connectivity, pride, and a sense of location. However, host cities, regions, and countries face a number of challenges, including financing, effective governance, and the ability to accurately assess the economic, social, environmental, and other added value of tourism events.

Event management is a branch of project management that focuses on the planning, creation, and execution of events. It is the preparation and administration of an event or operation. It entails researching the complexities of the brand, defining the target audience, developing the event model, organizing the logistics, and managing the logistical aspects before actually carrying out the moralities of the planned event. Good interpersonal communication, negotiation, budgeting, and innovative skills are needed for event management.

There is no widely accepted concept of an event. Many scholars have written about the meaning of events and the different words that are used to define them. However, there is only minimal agreement among the various studies on standardized terms.

According to Philip Kotler, "events are defined as occurrences to communicate particular messages to target audience."
A comprehensive multimedia definition states that an event is a multimedia package carried out with a preconceived concept, customized or modified to achieve the client's objective of reaching out and suitably influencing the sharply defined and specially gathered target audience by providing a complete sensual experience and an avenue for two way interaction.
Bowdin (2006) notes that the term —event has been used —to describe specific rituals, presentations, performances, or celebrations that are consciously planned and created to mark special occasions and/or to achieve particular social, cultural or corporate goals and objectives.
I Jago and Shaw (1998) suggest six features of special events. According to them, special events should: Attract tourists or tourism development, Be of limited duration, Be one-off or infrequent occurrence, Raise the awareness, image, or profile of a region, Offer a social experience, Be out

3.3.1. Classification of Events

There are several standards for categorizing events. The most basic distinguishes between planned and unplanned events. Planned events are the focus of event management research, and they necessitate setup, management, executives, and a fixed amount of time. Accidents, natural disasters, and other similar events are examples of unplanned events that will not be considered in this chapter.

Events have a distinct type of tourist attraction, varying in size from small cultural festivals to international trade fairs to the biggest global sporting events, including the Olympic Games and the FIFA Football World Cup. One of the primary distinctions between activities and typical attractions is the time span during which they have an effect on the host community or country. By definition, events are short-term, lasting just one or two days, though some larger events may last much longer, while fixed attractions draw visitors seasonally or over a long period of time. The main factors that are commonly recognized as defining the perceived size and effect of events are the level of attendance, audience/spectators, and media attention, as well as the degree to which an event creates substantial international demand for each.

3.3.1.1. Mega Events

Major or mega-events are traditionally described as having international appeal and true global scope. Such events have the potential to serve as catalysts for local growth and to provide a variety of economic, socio-cultural, environmental, and other benefits related to image, branding, and visitor economy expansion, to name a few. Mega-events, being the largest and most visible of all events, inevitably necessitate the most significant and sophisticated infrastructure construction, are usually the most costly to host, and take the longest time from conception to delivery due to the competitive bidding process for such events. They still have the most time to leave a legacy. However, there are very few chances for cities and countries to host these mega-events. Some of the limitations in the organization of these events include concerns with infrastructure, facilities, transportation, and cross-cultural issues. Despite this, many countries continue to see the expenditure of capital required to compete for and eventually host these mega-events as one that can yield a return on investment (ROI). The specific aims and desired outcomes from a national perspective which differ between potential host candidates for the largest of events, depending on their level of development, existing infrastructure, identified security needs, plans for urban regeneration, existing, and desired international image, maturity as a tourism destination, and sustainability credibility, to name a few. As a result, it is clear that different events have different possible legacies, depending on the needs of the host city or country. Getz (2005) describes them as follows: Mega-events are those that, due to their size or importance, generate exceptionally high levels of tourism, media attention, prestige, or economic impact for the host city, location, or organization. Another author,

Hall, explains that mega-events get their name from their size in terms of attendance, target market, extent of public financial participation, political consequences, extent of media coverage, facility construction, and influence on the host community's economic and social fabric (Allen et al., 2011).

3.3.1.2. Regional Events

Regional events are intended to boost thc popularity of a particular tourist destination or country. FAN fair, the world's largest country music festival, held annually in Nashville, Tennessee, the Kent yucky Derby, and Arts in the Park, a Memphis art Festival are all examples of tourist destinations achieving market positioning for both domestic and foreign tourism markets through annual events. Another example of a regional event is the annual National Chewy Blossom Festival in Washington, D.C.

3.3.1.3. Major Events

These events generate considerable local awareness as well as a large number of participants, resulting in significant tourism revenue. Chinese New Year celebrations, for example, are held in several capital cities. In Honolulu, the celebration involves several New Year's festivals and rituals, such as the Lion Dance, lantern festivals, parades, and dragon boat races. Friends and relatives from the Chinese community often visit during this period. Most major cities have a convention center that can accommodate large gatherings; trade is recognized as North America's premier convention center, attracting over 4 million trade and public show visitors each year. The McCormick Place complex consists of state-of-the-art buildings with a total of more than 2.2 million square feet of exhibit space, 1.6 million square feet of which is all on one floor, making it the nation's largest convention center.

3.3.1.4. Minor Events

The majority of events fall into this last category, and it is here that most event managers gain experience. Annual events are held in almost every town, region, state, and world. There are literally thousands of agricultural fairs and expos around the country and state. There are also regional activities held each year, the biggest of which is the Texas state fair, which attracts over 3 million visitors each year. Aside from annual celebrations, there are a few one-time events, such as historical-cultural musical and dance performances. This category includes meetings, celebrations, festivals, conferences, award ceremonies, exhibition sporting activities, and many other cultural and social events. In terms of types, events may be categorized as follows also:

1. *Sporting:* These events are held in every town, city, and state throughout the country. They draw world-class athletes from all over the world. Tennis, golf, baseball, football, basketball, downhill skiing, and car racing are a few examples. These major activities are matched at the local level by athletic tournaments for all levels of players. For example, the Program, which is held annually at the majority of golf courses, allows participants to play alongside professional golfers. This event is normally the highlight of the golfing calendar and necessitates a significant amount of effort on the part of the team that supports it, which includes the club committee, the club manager, the club professional, ground Staff, club administration, and catering (Figure 3.8).

Figure 3.8. Sport events.

Source: https://th.bing.com/th/id/OIP.mo3viLZQkHG_eFyoHwn5cwHaEK?pid=ImgDet&rs=1.

2. *Entertainment, Arts, and Culture:* The potential of entertainment events to draw large crowds is well recognized. In some cases, the concerts are highly financially viable; in others, financial issues will easily intensify when ticket sales fall short of expectations.

3. *Commercial Marketing and Promotional Events:* Promotional events usually have large budgets and a high profile. The aim of promotional events is to distinguish the product from its competitors and to make it memorable. The majority of promotional events include product launches, which are often for computer hardware or software, perfume, alcohol, or automobiles. One such marketing activity dazzled attendees by having its latest launch motorbikes ride overhead on a tight rope, complete with special effect lighting (Figure 3.9).

Figure 3.9. The aim of a promotional activity may be sales, such as travel agents promoting the tour to their clients or prospective purchasers.

Source: https://th.bing.com/th/id/Rd4203bf76c72f2da890be802d025eb8c?rik=61W6QqMDyfisNw&riu=http%3a%2f%2fwww.businesspromotions.net%2fwp-content%2fuploads%2f2018%2f10%2fSimplified-Sales-Promotion-Through-Personal-Campaigns.jpg&ehk=sksNDWA04dnllRI8T6YtrVsfd%2bXo%2bKyt14TlFYsyNag%3d&risl=&pid=ImgRaw.

Since the media is normally invited to these events, the impact, publicity, and risk are all high. Success becomes critical.

4. *Meetings and Exhibitions:* The meetings and exhibitions industry are a cutthroat one. Many conferences draw thousands of people, while some meetings draw only a few high-profile attendees.
5. *Religious Festivals:* This group includes all religious festivals. Wine and food festivals, as well as harvest festivals, are becoming increasingly common, allowing a specific region to showcase 15 cases of its commodity. Chinese New Year and harvest festivals in India, such as Pongal in Tamil Nadu and Onam in Kerala, are good examples (Figure 3.10).

Figure 3.10. Religious festivals events.

Source: https://static-blog.treebo.com/wp-content/uploads/2018/02/christmas.jpg.

6. *Families:* Weddings, anniversaries, birthday parties, and even funerals all offer ways for families to come together. Asian visitors represent a sizable market for the wedding industry, with many couples opting for a traditional ceremony at home. It is important for the Event Planner to stay on top of these shifting social patterns.
7. *Fundraising:* Fairs are popular in most communities and are often organized by enthusiastic local committees. The amount of effort

and organization needed for these events is often underestimated. Since their overarching goal is to raise funds, there is a possibility that participants will invest all of their money on these events while ignoring those that are more profitable to the charitable cause. The charitable fundraiser must comply with a variety of legal standards.

3.3.2. Benefits of Events

From the standpoint of the visitor economy, the possible advantages of hosting major events include:

1. ***Structural Expansion of the Visitor Economy:*** Visitors coming to a city or area for an event will help to boost the economy, as tourist spending has a multiplier impact on incomes in associated supply chains. The host destination would benefit from the multiplier effect in terms of employment, wages, and higher living standards.
2. ***Tourism's Alignment with Other Strategies:*** The standards of hosting a major event can be used to encourage an integrated whole-of-government approach and optimize synergies between relevant development and growth infrastructures built for events are one of the most noticeable long-term legacies for a host city or area and can have real impacts on tourism growth.
3. ***Marketing and Promotion:*** Pre-event branding associated with the effective hosting of a major event will provide long-term awareness of destination branding in key tourism markets, promote repeat visitation of attendees or participants, and provide a greater understanding of the event's focus, such as sport, arts, and culture, food, and wine, and so on.
4. ***Environmental Impacts:*** The international emphasis that is frequently associated with major events will assist in prioritizing work on an often underdeveloped or overlooked built environment, and thus the attractiveness and competitiveness of destinations. Furthermore, ensuring that events are handled in an environmentally sustainable manner is becoming a top priority in terms of branding. A positive legacy can inspire community and stakeholder support for an event, represent a measurable ROI, or serve as justification for government spending. To achieve a positive outcome, however, strategic planning well in advance of

the event is needed, as is the adoption of a long-term outlook and assessment over the event lifecycle, from inception to post-event. Any infrastructure planning must prioritize long-term usage, preferably with funding and resources earmarked for the post-game's era.

3.3.3. Need for Event Management

1. ***Provides Career Opportunities:*** Career prospects in event management include event planners, event managers, operations, and logistics managers, entertainment managers, security managers, and so on.
2. ***Professional Growth Opportunities:*** Event managers have a variety of potential for advancement if they continue their education and work on various projects to obtain all-around experience. Event planners/managers may become entrepreneurs with their own event management companies or independent consultants with time and experience.
3. ***Opportunity to Switch Careers:*** Event management is one of the few careers that offers a great deal of opportunities to break into the field quickly. It does not necessitate a formal degree or other professional qualification. Professionals in other fields will quickly transition to event management if they so choose.
4. ***The Availability of More Work Opportunities:*** Event management is a discipline that is expanding at a rapid pace and creating more opportunities than ever before.
5. ***Better Pay Packages:*** In comparison to many other professions, the income prospects in event management are among the highest. Pay packages improve over time and practice.
6. ***Global Reach:*** Event management is a career that can take people all over the world. The type of experiences event management delivers in planning and executing events on a worldwide scale is immense. The scope is global in nature, whether it is the management of an event catering to international audiences or the execution of an event in a foreign location.

3.3.4. Objectives of Event Management

Every event must have a specific overarching goal; otherwise, the event does not take place. Events necessitate a significant amount of concerted effort and dedication. This dedication can only come from a sincere conviction among all participants that the objectives are worthwhile and will be successful in the long run. In addition to an overarching objective, each individual event must have its own list of objectives, which must be transparent and written down in a way that allows you to measure the event's progress after it is completed.

Objectives should always be SMART:

- **Specific** to the certain event and exact aspects of it;
- **Measurable** demonstrates the objectives in numbers and quantities;
- **Attainable** make certain that all team members know the objectives;
- **Relevant** set objectives the organizing team can realistically attain;
- **Time-Bound** set a span of time for achievement of the objectives (Figure 3.11).

Figure 3.11. Objectives of event management.

Source: https://files.nc.gov/ncoshr/styles/main-image/public/SmartGoals.jpg?itok=e5ErJ0R6.

3.4. FIVE C'S OF EVENT MANAGEMENT

1. ***Conceptualizing:*** This is the first "C" in event management. The event planner learns the precise purpose for which an event is being prepared from the client. This allows the team to develop

a fitting theme or idea for the event's centerpiece. The theme should be developed with the event's purpose, target audience, preliminary location, media to be used, and so on in mind.

2. ***Costing:*** This entails the development of budgets, cost forecasts, and the sources of funding. It is often done to make a provision for risk coverage and to manage uncertainties.
3. **Canvassing:** These activities include networking and advertisement to create the necessary mileage for the event's success, with the ultimate goal of raising funds and securing sponsorships for the event's execution.
4. ***Customizing:*** The customization of an event is based on the client's satisfaction. Reinforcing client requirements toward objective fulfillment and maintaining a balance of innovation and suitability to changing client patterns and preferences. Involves a customized strategy, tailored specifically for an event, in order to be effective in reaching the target audience and meeting the client's needs.
5. ***Carrying Out, Culmination, and Control:*** This is the process in which plans are actually carried out. It encompasses all of the events and operations that contribute to event execution and achievement of the event's goal. In fact, each of the Cs cannot follow the order in which they were presented above. Before the Carry out point, there is a dynamic interaction between the various C's, depending on client requirements, budget constraints, and so on. The original concept is modified and revised based on the feedback received from the other C's (Figure 3.12).

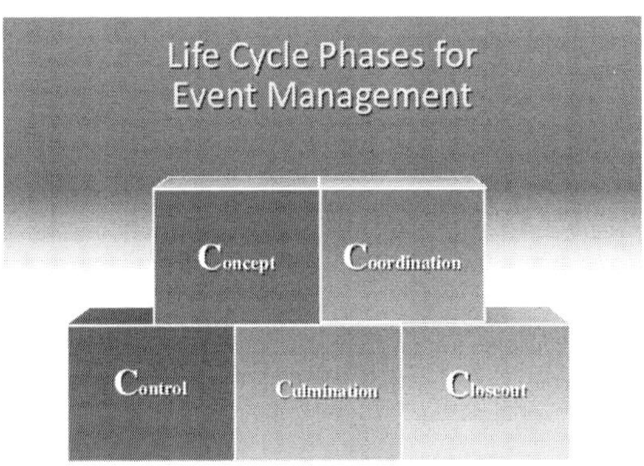

Figure 3.12. Life cycle phases for event management.

Source: https://th.bing.com/th/id/Rd13d6ed0357bd9474bf9d9c49f4e765a?rik=DztrZIt9lIQw2A&riu=http%3a%2f%2fimage.slidesharecdn.com%2fthe5csofeventmanagement-130821081155-phpapp01%2f95%2fthe-5-cs-of-event-management-3-638.jpg%3fcb%3d1377072792&ehk=OQzbfgehhQ1Ai3gdSK642pUpMhY9VZJWkon%2fbF3HRiE%3d&risl=&pid=ImgRaw.

3.5. ANALYSIS OF EVENTS

- ***Step 1: Select Events for Analysis:*** The event manager must first choose the desired set of events and then decide which event to analyze first. If there have been several events, he must rank them in order of importance. If the events are of a similar nature, the event manager can combine them; however, if the event is very serious, the event manager must appoint a special team to conduct a thorough analysis of that particular event.
- ***Step 2: Collect and Collate Information:*** After deciding which event to analyze, the event manager must thoroughly collect and collate all information related to that event in order to begin planning for that event.
- ***Step 3: Organize the Analysis Meeting:*** The event manager must also ensure that regular meetings are held prior to the event's occurrence. He can also invite outside participants who may have more information about the event for the meeting. All such participants can also make specific suggestions for how to improve the event.
- ***Step 4: Set Frameworks and Rules for the Analysis:*** When a meeting is held for the purpose of analysis, the event manager must make certain rules for his meeting. The rules he establishes must be consistent with the event's main goal. Furthermore, the meeting's discussion must be primarily focused on achieving these objectives.
- ***Step 5: Carry Out the Analysis:*** The event manager then conducts the analysis to determine if there are any problems and what kinds of initiatives can be made to ensure them.
- ***Step 6: Implement Changes and Follow Up:*** If the event manager discovers during the meeting that changes are required, he can execute those changes and follow up on them.

3.5.1. SWOT Analysis in Event Management

SWOT analysis refers to a strategic planning tool used to recognize and investigate the strengths, weaknesses, opportunities, and threats involved in events. SWOT analysis can also be done on any organization (Table 3.1).

Table 3.1. SWOT Analysis

Internal Environment	**External Environment**
Strengths	**Opportunities**
Highly motivated staff Customer Execute plans within budgets Interactive activities	Growing demand New markets New acquisitions New products and services Support from local authorities
Weaknesses	**Threats**
Tax structure Brand portfolio Future profitability	Rising cost of raw material Poor infrastructure No derived license/special permits

1. ***Strengths:*** These are the characteristics of the project or organization that help it achieve its goals. For example: an experienced event team, high motivation levels, excellent PR, a large market share, and so on.
2. ***Weaknesses:*** These are the characteristics of the project or organization that are detrimental to achieving the project's objectives. For example, social loafing, a lack of resources, an unskilled event team, a lack of energy, an absence of media and corporate contacts, and so on.
3. ***Opportunities:*** These are external factors that aid in the achievement of project objectives. For example, low competition, favorable economic conditions, local government support, availability of cutting-edge infrastructure, and so on.
4. ***Threats:*** These are external factors that are detrimental to achieving the project's objectives. For example, high competition, little or no support from local authorities, bad weather, poor infrastructure, high labor rate, and raw material scarcity. It is critical that we conduct a SWOT analysis before developing an event plan in order to develop a strategy that maximizes

the potential of the project's strengths and opportunities while minimizing the impact of the project's weaknesses and threats.

3.6. KEY DECISION MAKERS

1. ***Event Manager:*** This is in charge of ensuring that the event runs smoothly. His decision-making capacity is shown by the fact that he addresses the event's goals, priorities, and other essentials with the client and schedules the events accordingly.
2. ***Client:*** The client asks the event planner about organizing his preferred event. The client has preconceived notions. The client communicates his suggestions to the event planner, who turns them into events. The Client expresses his hopes for the case.
3. ***Venue Manager:*** The person who is in charge of the venue, which may be a hall, theater, conference center, or hotel, is known as a venue manager. Their primary responsibility is to supervise operations and facility use, which includes ensuring the venue is clean and all equipment is operational. Bookings for viewing the stage, bumping in, and out must be made through the venue management team.
4. ***Catering Manager:*** These prepare, coordinate, and oversee the organizations and business's food and beverage facilities, ensuring that consumer needs, food, and sanitation requirements, and financial goals.
5. ***Logistics Manager:*** These are in charge of the movement, delivery, and storage of merchandise, instruments, and equipment at an event site. Managers are in charge of hiring, training, and evaluating staff. They establish worker schedules and ensure that staff adhere to safety regulations.
6. ***Government:*** Obtaining approval from the government is needed for some activities, such as foreign or religious events. International events generate revenue for the country. Permission and facilitation from the government for events is an important feature of event success.
7. ***Security Managers*:** These are in charge of maintaining law and order at the venue. Important decisions relating to the upheaval encountered at previous events can be an important factor when making security arrangements.

8. ***Hospitality Manager:*** The primary responsibility of staff would be to communicate with customers and meet their needs. At meetings, the hospitality team uses their public relations expertise to assist customers.

9. ***Media Managers:*** The production of media is the bundling of internally or externally generated content and its conversion into a medium. At events, the media plays a major role. Creating media content is a significant task.

3.7. FUNCTIONS OF EVENT MANAGEMENT

1. ***Planning:*** It is a process to improve resource utilization across the board. Given the complexities of decision-making and the need for various event activities, a cross-functional team is required. Beginning with an understanding of the client profile, the event brief, the target audience, and the number expected, a major component of any event that follows is the preparation of the event budget. The planning function is engaged in micro-level event agreements such as liaising with the creative team to discuss, facilitate, and arrange technical specifications such as sound, light, stages, and sets. One of the most difficult tasks in the planning function is shortlisting artists and standby artists in accordance with the dictates of the creative artists. It also entails investigating alternative arrangements for locating the event, the venue, the event's conditions, and gathering information to aid in deciding whether the event should be held indoors or outdoors. While on the subject of the last task, understanding the requirements for licenses, clearances, and so on and arranging for them as and when necessary is a fundamentally responsible task that the event coordinator is tasked with. Soft tissues such as whether the show will be ticketed, nonticketed, fully or partially sponsored are also considered during the planning process. The event is then rated in terms of risk by event planners. Defining structures for the quality of hospitality and the dress code of the hosts/hostesses depending on the audience profile, as well as deciding on the appropriate food and beverages to be served on the occasion, are all planning functions. This is especially true because the security and other arrangements will differ depending on the type of beverages served. Inflows to the event company

are primarily derived from a combination of sponsorships, ticket sales, commissions, event production charges, artist management fees, and infrastructure and equipment rental charges, as shown in the cash flow statement. All of these flows should occur concurrently; on the outflow side, one can include headings such as supplier payments, venue hiring charges, payment to artists and performers, and so on. The major outflows, however, are primarily related to event production, as well as licensing and tax payment liabilities. The mode of payment for events ranges from part payments to cash payments and is discussed and agreed upon by the parties involved before being authenticated in the form of a contract after negotiations. Penalty clauses for payment defaults may also be included. The actual payment plan can be worked out depending on the nature of the project, the relationship with the clients, and the objectives of both the client and the organizer. This may entail a certain amount as a down payment in advance, a certain amount upon completion of specific milestones, and finally payment of the balance amount either at the start or at the end of the event. It is critical that a certain amount be set aside as advance to cover working capital requirements. The planning function establishes the boundaries of the creative function by supplying the constraints that the creative team must work within. It deals with hard practical realities such as logistics (transportation of materials, travel, stay, etc.), and networking. It attempts to create a perfect picture of the event flow and to define and control the inflow and outflow of money prior to, during, and after the event. As a result, it is critical that the planning function play a significant role in the preparation for any event. Furthermore, because the time frame for decision making is limited, planning becomes an even more important function. Travel arrangements, audiovisual needs, catering, China, and flatware, convention services, decor, decorations, and props, entertainment, exhibitor needs, site selection, sound, and lights, speakers, stage decor, staging, web site management, and so on are some of the event planning services that must be handled by event organizers.

2. ***Organizing:*** These events typically have a team-based work atmosphere and a project-type organizational structure, and responsibility for the event is delegated to the appropriate staff

members of the team. Coordination of the required arrangements is divided among the team members. Understanding planning in the sense of event management entails describing the activities needed for an event, defining individual and team roles, and allocating responsibilities to coordinators. A consistent delineation of authorities and delegation of authority are also part of the procedure. This type of exercise aids in the development of an intentional framework for the clarification of roles and positions. Depending on the resources available, these systems change with almost every event. In event management, project-based structures are more common. Event coordinators are primarily responsible for event organization. One of the most critical duties of the event planner is to contact the artist or performers and, in the event of absence or dropout, to make standby arrangements. After the organizing and creative functions have worked out the game plan, the event planner works with the artist to finalize the date, terms, and conditions. This is accompanied by the development and design of the requisite infrastructure. Following the planning and coordination with professionals for the physical availability of the sound, lighting, stage, sets, and seating, some softer aspects of organizing are arranged. This includes managing marketing, which includes press meetings, announcements, and so on, as well as handling ticketing and invitations. Once the specifications are determined during the planning stage, the coordinator is responsible for obtaining permits and licenses from different government agencies. Organizing functions include arranging for hospitality management such as lodging, food, and drinks, hostesses, and so on, as well as contacting sponsors to ensure the fulfillment of promises made by the event manager to their clients. In a nutshell, organizing is the process of making an event happen within the constraints established by planning.

3. ***Staffing:*** Event management staffing specifications are defined by functional roles in a project-type organizational framework. In event management, the importance of team structure, experience, context, and knowledge of team members is critical. The size and availability of resources in the events enterprise, to some degree, determine the exact position of the staff members. Event management necessitates the use of personnel with a variety of skills to oversee a wide range of activities. In a large firm, there is more

space for specialist functional personnel with minimal functional duties, while in a small firm, positions are merged based solely on time and staff availability. Thus, when interviewing for events, it is common to believe that candidates with prior experience in the hospitality industry, sales, and advertisement will be best suited to cope with the tension and uncertainty that the process entails. As previously mentioned, the events are very physical in nature. A large number of professional and unskilled volunteers and labor workers must be efficiently guided. Functionally, the following functional level tasks that must be handled within the team for a particular event, as mentioned above in the section on organizing, may be segregated. The overall organizer is in charge of a specific event. He has the final authority in event-related decision-making. The creative manager is in charge of the creative team. The project manager's job is to ensure that the event is conceptually successful, and he or she plays a critical role in the planning function. The production managers are active in the planning stage as well, but their primary duty is to ensure the event's physical performance.

4. ***Leading and Coordination:*** The number and substance of events revolves around interpersonal abilities as a whole. The primary objective of teamwork is to achieve unity among individual efforts in order to achieve the team goal. The overall coordinators must be administrators with exceptional people skills. Given the physical nature of the task, the time constraints involved, and the one-time nature of the event, they are constantly forced to inspire the workers and other junior coordinators to work extremely hard. Given the scarcity of experienced event planners, the overall organizer should also be able to direct the marketing and project managers. This could also imply that the knowledge and expertise of previous events must be passed on to relative newcomers. Thus, excellent communication skills, patience without allowing too many mistakes, and understanding how to use the carrot and stick in a neutral way are the basic characteristics of the overall coordinator. In addition to the above, the ability to identify and manage problems is a desirable management quality for an event planner.

5. ***Controlling:*** The essence of controlling is the evaluation and correction of deviations in event plans to ensure compliance with

original plans. Evaluation is a process that attempts to comprehend and quantify the degree to which an occurrence has succeeded in accomplishing its goals. The object of an event varies depending on the category and variety of event. There are two methods to putting assessment into proper context. The evaluation definition mentioned above was a critical review to determine what went wrong. Making suggestions on how an event should be changed to accomplish its goals more efficiently is a more positive focus for assessment. It is important to understand the predefined goals of the activities before conducting an assessment and measurement exercise. The brief should provide all of the information that will be shared, and if an event is planned without a clearly specified objective, any assessment of it will be meaningless. The basic event assessment process consists of three steps: defining tangible goals and integrating sensitivity in evaluation; assessing performance before, during, and after the event; and finally, correcting deviations from plans.

3.8. PRINCIPLES FOR HOLDING AN EVENT

1. ***Principles of Clarity of Purpose (Clear Objectives):*** The principle of clarity of purpose specifies that the company's goals should be stated clearly. These goals may be financial or operational in nature. The financial goal could be to achieve the desired ROI. If the event management team has agreed on the goals, the event can be organized accordingly. The difficulty for an event management company is that each event is a separate project for the company. Before the event plan starts, the aim should explicitly state observable outcomes. Different events necessitate a variety of objective and practical objectives. Based on different incidents, the company's goals should be modified, and it should hit the ground running.
2. ***Principle of Focus (Target Audience):*** There is no business without customers. Customers are a company's most valuable asset. As a result, the second event management concept is customer attention. Any event hosted by the organization should put the consumer first. Companies can schedule activities and then invite the appropriate crowd, or they can create events specifically for a specific audience group. It is best to start with the customer

in mind and plan the event from there. To understand customer needs, the organization should perform market research on a regular basis. The business will determine their target audience and the services they need by conducting market research and understanding what their competitors are doing.

3. ***Principle of Communication (Media Contacts):*** The organization should be aware of the organization's objectives and goals. It is also important to explain to the outside world what the organization is doing and why it is doing it. The company should make every attempt to reach out to the outside world and communicate its presence and the value of its existence in the market. The business can communicate to the outside world the value it adds to society by providing different services. Maintaining good public relations is a must for any event management organization. Event management firms may use social media to establish their presence in the market. They can easily use these media to spread their ideas to the general public.

4. ***Principle of Connectivity (Logistics):*** Events of the company may be organized indoor or outdoor. The location is chosen based on the type of events that the organization has planned. If the activities are held outside, it is important for the organization to provide adequate logistical facilities for the audience. There should be no ambiguity about the mode of transportation to the site. Participants should be informed about the logistics aspects at the time of registration. Participants should be aware of the pick-up and drop-off locations. In addition, the company should ensure that the required men and materials are transported to the location well in advance to promote the desired preparations.

5. ***Principle of Facilitation (Engaging the Audience):*** Because of the audience, the company earns sales. It is the company's primary responsibility and concern to involve the audience and provide them with a joyful experience. This is only possible if the organization has secured a high-profile speaker and a strong organizing team. These team members' talents captivate the audience. The audience will be kept entertained by entertainment activities, sports, and competition. These are the activities that linger in the minds of the viewer and entice them to return to the company's other events. The primary goal should not be overlooked when carrying out these tasks.

6. ***Principle of Promotion (Giveaways):*** According to the promotion principle, an organization should aim to support itself in all of its activities. It is a good idea for the organization to have some kind of gift to the participants. This is done to express gratitude for their sponsorship of the event. The nature of the gifts should not tarnish the company's reputation. Depending on the nature of the business, it could be possible to offer away an existing product. Using an existing commodity as a gift promotes it directly. If the company's own goods are used, there is no need to spend extra money on other gifts for obedience. Nothing should be given as a gift that contradicts the event's very basic intent.
7. ***Principle of Confidence (Ready to Take Challenging Questions):*** People are inquisitive listeners and learners. They would want to look into every possibility in order to learn more about the event or the business. As a result, the organization should be prepared ahead of time for any questions from the audience or any of the participants. Not only should the individual provide compelling answers, but he should also provide accurate answers.
8. ***Principle of People Management (Providing Proper Staff):*** The success of any event is largely dependent on how good the event was organized and coordinated. As a result, it is important for the organization to have the right employees for the right job. The event management team should assign various tasks to a responsible team leader. The team members must be adaptable enough to deal with the event. They should be willing to give up their personal lives. This is valid if the event is held on a weekend. An employee who is dissatisfied with his or her job cannot provide satisfaction to his or her client.
9. ***Principle of Documentation (Recording the Event):*** The organization should design and implement a proper documentation mechanism. The device should be able to record all events in great detail. The organization must be able to keep all of the documents in a very systematic manner. The framework should be very versatile, in the sense that it should be capable of producing any report in any format. For this reason, the organization should make use of existing technologies. There is a wide range of counting and record keeping tools available to help you complete this task with ease and precision. These documents may be used to assess a variety of factors. Records assist the organization in

determining the number of participants participating against the number of participants predicted. The reports reveal the total sum spent as well as the budgeted spending, budgeted revenue and actual revenue received, and so on. These documents will help the organization in future event planning.

10. ***Principle of Evaluation (Review):*** Any activity's phase will end with evaluation. "If you can't weigh it, you can't handle it," goes the adage. As a result, an event planner must assess each event after it has concluded. As previously mentioned, it is the duty of the event management team to establish the assessment target for each operation. The actual result should be compared to the pre-determined criteria. Both positive and negative aspects should be highlighted in the evaluation. Positives should be recognized, and the organization should try to eliminate vulnerabilities. Event management should view assessment as a continuous process. A routine evaluation would increase the efficiency of the company's activities. The objective assessment of all operations at regular intervals ensures the smooth operation of the entire case.

11. ***Principle of Customer Convenience (Easy Registration and Access to Event Place):*** Customers' convenience should be the primary concern of every organization. Customers want to have a positive experience with the brand. The quality of the organization is heavily reliant on how the company treats its customers. Customer convenience and comfort can increase customer loyalty. As a result, the company should make every attempt to keep their customers satisfied and pleased. This should happen right away, during the registration process. Customers should have simple access to registration, follow-up, and payment. The organization will use technology to help it in this direction.

3.9. THE 10 EVENT MANAGEMENT SKILLS EVENT MANAGERS NEED

The organization of any large or small event, ranging from personal events such as weddings and celebrations to business-to-consumer (B2C) events such as festivals and business-to-business (B2B) events such as conferences and exhibitions, is referred to as event management.

There are several parallels in event management. A tradeshow, for example, maybe targeted at companies, such as the Mobile World Congress, which is held in Barcelona each year, or can be aimed at consumers, such as the Ideal Home Show, which is held in London.

Event management may refer to any aspect of the organization and management of the event, such as setting goals, selecting vendors, creating timelines, and handling the event on the day of the event. There are many moving parts in every event, and only one opportunity to get them right, so good event management skills are important for success.

3.9.1. What Does an Event Manager Do?

An event manager/planner is the person in charge of ensuring that an event, wedding, meeting, or festival runs smoothly and that all of the event goals are met. They are in charge of everything from ensuring that vendors arrive on time and that the presenter schedule is booked to ensuring that the event adheres to health and safety or environmental issues.

An event planner often operates on behalf of a brand and is the main person responsible for ensuring that the event runs smoothly, while an event organizer is frequently the individual who determines whether or not to hold an event. An event coordinator will need to recognize gaps in the events market, study competitor events, plan new events, and maintain a calendar of events throughout the year, while most of these tasks may have been completed by the time an event planner is hired. Event organizers can work for a brand or for an event organization company that specializes in event management. An event organizer can employ event managers or have others on their team who help to ensure the success of the event.

Event coordinators help to ensure the smooth operation of an event and maybe in charge of one or more aspects of the event planning process. Some event coordinators, for example, specialize in event management such as registration or speaker liaison, while others handle broader duties such as design and styling.

3.9.2. What Is the Role of an Event Management Company?

In most cases, event management entails overseeing the entire process, from start to finish. An event management company is often employed by a brand, another agency (such as a marketing or advertising agency), a location, or even a destination, such as a country's tourism board.

Event management firms are typically hired for their extensive network of connections, their expertise, and their capacity to make an event run smoothly. While the end customer, such as a brand or a destination, may know what they want from an event, they may not know how to get there. This is where an event production company with a team of professional event planners will come in handy.

An event management company may be in charge of the following areas:

- Creating an event budget;
- Sourcing of venues;
- Managing suppliers for instance: lighting, caterers, audiovisual, health, and safety, and event staffing;
- Styling and design;
- Organizing marketing as well as promotion;
- Reviewing and risk management;
- Managing onsite deployment and event success;
- Post-event analysis.

3.9.3. What an Event Management Proposal Might Look Like?

When a corporation or organization wants to employ an event management company, they may frequently request an event management proposal. This is an opportunity for the event management company to demonstrate their abilities, prior experience, and what they can provide for the potential event.

As an example of an event management proposal, consider the following:

1. ***Company Introduction:*** This is the opportunity to have a perspective on your company's past, team, and specialties.
2. ***Previous Experience:*** In this section, you will share previous events you have served on that are similar to this one, as well as client testimonials and references.
3. ***Specialty Areas:*** This is the part where you can share your knowledge. Do you plan events from start to finish, or do you focus on a specific feature, such as attracting great speakers or organizing marketing activities?
4. ***Event Proposal:*** In this section, you can answer the client's brief as simply as possible. To ensure that you address and answer all areas, it is a good idea to mimic the layout or headings of the brief you have been given.
5. ***Pricing Schedule:*** This may be a spreadsheet or a costing page where the budgets from each segment are combined with a final

number. This may also provide information about your pricing terms.

6. ***Timelines:*** Include any deliverables and provide a schedule on what needs to be done and when in order for the event to be a success.
7. ***Conclusion:*** This is a brief overview that normally includes a thank you and can also contain a portion for all parties to sign if they agree.

3.9.4. Top 10 Skills Needed for Successful Event Management

There are a few skills you must master in order to be a successful event manager. Here are some of the abilities that good event planners would have:

1. ***People Skills:*** What is the most important characteristic shared by active event managers? People skills. You must be at ease communicating with high-level executives, government leaders, vendors, coworkers, sponsor representatives, clients, managers, suppliers, employees, and event attendees. To work successfully with such a diverse group of people, you must be able to resolve conflicts quickly, be a confident but pleasant negotiator, and maintain your sense of humor.

   Remember to enjoy your work and the people you work with, and they will want to work with you again. You cannot do it all by yourself, so developing relationships is important.
2. ***Flexibility:*** Event managers must double up as housekeepers one minute and fearless leaders the next. Cleaning up spills and making difficult decisions — easily, quietly, and effectively — are all part of the work. Maintain you are cool, get the job done, and then get back to running the show.
3. ***Listening:*** Understanding what key stakeholders expect from the event is important. These people may not be in the event business, so they may not understand the jargon or know what is realistic. You must be able to identify their desires and ensure that both parties have the same goals. Pay close attention to what is said- and what is not said-during key conversations. Using these spoken (and unspoken) needs to guide your preparation will keep you one step ahead.

4. ***Organized:*** To run a successful event, you must be able to juggle around 50 items at once. This ability to multitask is needed for event planning as well as flawless execution. The most successful planners have foolproof systems, step-by-step checklists, and useful technology.

   Working in events necessitates the ability to see the larger picture when paying attention to the smallest information. Get used to delegating some of the more time-consuming duties to prevent burnout. And, if things do not go as planned, do not be afraid to turn to plan "B."

5. ***Passion:*** Genuine enthusiasm enables you to conquer roadblocks and maintain your cool when all seems lost. It also results in bursts of creativity and the inspiration to make something amazing rather than just getting through the day. Things like time management can be taught, but passion cannot be taught, and you will need it to succeed. Being an event planner can be a thankless role at times; it should be about more than just a daily paycheck for you.

6. ***Communication:*** Simple, firm, and kind communication identifies you as the team's leader, keeps everyone on track, and ensures that everyone involved understands the event's objectives. It also enables you to easily express your vision and enthuse others about it.

   Communicate in a manner that is respectful of others and does not diminish anyone. Accept criticism while being open to fresh ideas. Everyone contributes to an event's success, so make sure you are engaging with them in a straightforward, optimistic, and inspiring manner.

7. ***Remain Calm in Stressful Situations:*** Your team will look to you for responses to almost everything. The last thing they need in times of tension is a weak leader who makes bad decisions because they are cracking under pressure. Even in the face of adversity, effective event planners maintain their cool and treat everyone with dignity. Even if you are a nervous wreck on the inside, try to keep your composure, calm, and collected when dealing with others.

8. ***Problem-Solving Abilities:*** You have to be resourceful about what you have, whether it is hunting down some emergency duct tape

or reworking a keynote presentation at the last minute. Everything will go wrong no matter how good you intend. And it will be a lot easier to solve if you enjoy coming up with innovative solutions to problems.

9. ***Conclusive:*** Event managers must be able to make many decisions at once and quickly. And, even more importantly, they must be able to understand when it is too late to reverse certain decisions — and have the determination to stick to them.
10. ***Experience:*** If you have a lot of experience, it would be much easier for people to believe you. And, now that you have seen the variety of obstacles that an occurrence can throw at you, your job will be easier. In addition, when things get difficult, the seasoned manager has a network of trustworthy professionals to call on. Do not worry if you do not have a lot of experience yet; just give it time!

Chapter 4

Sport Event Impacts

CONTENTS

4.1. INTRODUCTION

Tourism is a hospitality business, and visitors expect courteous and prompt service if they visit any attraction. In this regard, if tourism is to live up to its moniker as a hospitality industry, it must look beyond its own doors and employees to understand the social, cultural, and economic impacts it has on the host community (Murphy, 1985). As a result, it is important that tourism studies investigate the social impacts of tourism on host communities. Sport tourism, according to Hritz and Ross (2010), is one of the fastest-growing market segments in the tourism industry and is gaining recognition for its social, environmental, and economic impacts on destinations. Aside from that, prior research on tourism impacts has appeared to concentrate solely on tourism as a whole, with little distinction made between the various forms of tourism that may be present in a destination. According to Brunt and Courtney (1999), the social and cultural effects of tourism should be addressed in the planning phase and in an environmental impact assessment protocol to maximize benefits and mitigate problems. The engagement of communities in order for people to recognize tourism is an effective general planning strategy to reinforce positive and minimize negative impacts. This can be accomplished by including the group in decision-making and distributing the industry's benefits fairly (Brunt and Courtney, 1999). Furthermore, residents' views and attitudes toward the effects of tourism are likely to be a significant planning and policy factor for the effective growth, promotion, and operation of current and future programs and projects. The goal of this chapter is to examine the scientific literature on tourism sociology as a subject in order to study the economic, social, and environmental impacts of tourism on societies and residents, as well as how residents perceived the benefits and costs of tourism developments in the local community.

4.2. SOCIOCULTURAL IMPACTS

Tourism is a socio-cultural phenomenon that has grown in importance as technology and communication, and transportation systems have advanced. However, an increase in the number of people moving around the world does not necessarily indicate success or increased tourist access, but it can mostly serve more immediate marketing interests.

Because tourism is regarded as a phenomenon, there is a sociological interest in studying it. Tourist practice is an educational, learning process that is established through the relationship between visitors and residents, as well as their cultural backgrounds. Several authors devote their studies to this field, and others attempt to understand the relationships between tourists and residents in the host region.

4.2.1. The Impact of Sports on Human Society

Sports have a special place in human society because they teach us how to make a difference in our lives on a daily basis. Integrity and morality are lessons learned by incorporating sports into one's life. Sports teach many important values that have the potential to reshape the foundation of human society. Participating in sports teaches one that there are no social, political, or ethnic barriers. Individuals benefit from sports because it encourages them to stay physically active.

Sports help to foster a mindset in society that no matter what, one can always overcome any obstacle by taking a positive approach to it. Sports are an excellent way to demonstrate to society that a person's socioeconomic status does not define them; rather, their compassion defines them. Sports also play an important role in boosting a country's economy. Sports play an important role in fostering a healthy community by bringing people from all walks of life together.

Here are a few examples of how sports affect human society:

1. ***Sports Foster a Wellness Community:*** Sports have a significant impact on people's lives because they help to promote a fitness culture. People who compete in sporting activities have good health and physical strength to be able to perform correctly, which encourages others to adopt a fitness culture. Sports promote an active and safe lifestyle.
2. ***Sports Abolish Social Differences:*** Sport is an excellent tool for eradicating social inequalities because it treats people from all walks of life equally. Everyone is treated equally and given equal opportunities to perform, regardless of wealth. Regardless of social differences, the spectators sit together and enjoy the game.
3. ***Sports Help to Strengthen Community Bonds:*** Sports are an excellent way to foster community ties. Professional athletes have a community relations department, which means they also do community service. They organize campaigns and collect funds to assist those in need.
4. ***Sports Promote Safe Socialization:*** A person's mental health benefits from healthy socialization. When an individual participates in sports, he develops a more outgoing personality,

which provides him with many opportunities to develop healthy socialization with others.

Sports are important in everyone's lives, which is why they have such a large effect on human culture. Sports can help to change human society by creating positive opportunities for it to thrive.

4.3. POSITIVE AND NEGATIVE IMPACTS OF TOURISM ON CULTURE

4.3.1. The Positive Impacts of Tourism on Culture

Tourism has a positive impact on culture as well. According to Lantfant (1995), tourism, as a last resort, keeps uncultured or even closed cultural practices alive by recognizing, creating, rediscovering, and connecting them to the foreign market.

4.3.1.1. Cultural Preservation

Tourism provides support for the preservation and conservation of cultural heritage, restores cultural pride, revitalizes rituals and traditions, and opens the door to cultural exchange and learning (Smith, 2009). The majority of the popular positive impacts of tourism development on culture include increased cross-cultural contact, learning, preserving, and promoting local culture, arts, crafts, and traditions, encouraging host communities, and strengthening cultural values. Cole (2007) discovered that villagers in Indonesia are proud of tourism because it reinforces their cultural values. In order to draw distant visitors, children cultivate local traditions in order to build a solid and authentic foundation of cultural components without upsetting the values of their forefathers. Through photographing and promoting these identities in advertising materials at the provincial level, tourism has opened the door to establishing cultural identities such as homes, clothing, and megaliths dress as their collective identification and acceptance of political power. People claim that tourism has aided them in exploiting economic and political resources in order to obtain the power of church and a role in state administration. On the other hand, because of the increased social recognition and control, conflict, and crime could increase in major tourist destinations (Tsoudis, 2001).

4.3.1.2. Establishment of Institutional Infrastructure

Tourism undoubtedly raises awareness among locals who are not interested in their traditional arts and crafts industry. Tourism preserves these cultural elements by constructing certain institutional infrastructure (Witt, 1991). Many scholars are familiar with Deitch's case study of the Renaissance in Indian arts as a result of tourism (1989; cited in Page and Getz, 1997). Another example is given here. During 2006–2012, Counterpart International, a non-profit organization, preserved the strong 100-year-old Maya culture through the 'Community Tourism Alliance' project by co-managing and co-conserving approach. It has inspired women through the growth of entrepreneurship in various folk art and carvings, as well as pottering products. Later, it produced new promotional materials and broadcasted them around the world, as well as organized trade shows to draw visitors. The community is now claiming that they are proud of their culture because it can be seen on their crafts all over the world (Snyder, 2012). However, due to the nature of women's work, family conflict and divorce rates increased during this intervention (Nzama, 2008).

4.3.1.3. Derived Demand

Another strong example of localhost tradition preservation can be found in John M. Knox and Associates, Inc.'s (2004) case study in Hawaii, where the formal tourism industry extracted the demand for native Hawaiian culture in hotels. Following a meeting, representatives from the public and private sectors agreed that preserving and retaining native Hawaiian cultural themes would help the industry more than importing and manufacturing new components. As a result, there is a greater demand for native Hawaiian entrepreneurs who tell stories in their native language with honor and pride. Now, native Hawaiians claim that tourism allows them to show the world something important that can lead to human well-being. However, entrepreneurship brought wealth, strength, and the desire to earn more, which contributed to a shift in local lifestyles, increased crime, gambling, and alcoholism, and inspired cultural commercialization (Rátz, 2000).

4.3.1.4. Motivating Collectivism

People may shift from a self-centered to a collectivist mindset through social interaction. The city people of Jaipur were extremely self-centered, but thanks to tourism intervention, social interaction between people has increased in order to present a positive image to tourists. As a result, they learned about other countries and were able to overcome their prejudices about colonialist nations in India. This social interaction aided in demonstrating morality, fairness, and hospitality to strangers. This survey is done by Singla (2014). True, understanding between hosts and guests reduces misunderstandings and distance (Hunziker, 1961). Social interaction also contributes to the development of friendships between hosts and visitors, who spread positive word of mouth about hosts after returning home (Boissevian, 1979). Since the positive and negative impacts of tourism are determined by human factors, including the attitude and behavior of tourists and hosts during interactions, tourists, tourism, and hosts must share responsibility (Spanou, 2007). Instead of participating in the debate, seeking ways to manage the impact of tourism might be a good task. The overall socio-cultural impact management and the host-guest encounter points such as: goods and service purchasing, face-to-face contact for exchanging ideas and knowledge, and side by side at activities are essential management areas.

4.3.2. The Negative Impacts of Tourism on Culture

4.3.2.1. Acculturation Effects

Acculturation is the process of adapting and adjusting to a new environment (Ebin et al., 2001). When locals entertain visitors, they adjust to their needs, attitudes, and values and eventually begin to obey them (Nash, 1996). This is common in less sophisticated societies where the stronger community dominates the weaker (Burns, 1999). However, the diffusion of cultural component innovation in a social context is dependent on the compatibility, benefit, and complexity of the cultural artifacts (Lew, in press). Youth, in particular, are vulnerable to this effect, and women outperform their male counterparts in this regard (Ghuman, 1997). Following visitors' lifestyles causes changes in the material goods they use as well as their movements and postures, which Burns (1999) defines as demonstration effects. In the Pokhara-Ghandruk society in Nepal, the traditional fashion, behavior, and lifestyle of young Gurungs have been greatly affected by tourism (Pandy et al., 1995). Young people, for example, are wearing western modern clothing rather than traditional Nepali cloths and caps. They now favor western hairstyles, disobey their elders' Kinship title, engage in love-marriage rather than arrange marriage, drink with friends or host a dinner party outside the house, and ride bicycles rather than carts. The elders are vehemently opposed to and concerned about this cultural change. However, after careful examination of the case, it is discovered that not only tourism, but also other factors such as easy access to Indian movies, clothing, and communication vehicles, contribute to this effect. Tourism should be praised in this country because tourists provided ideas, education, and modern facilities that helped young people break free from superstition and find freedom in their work and relationships with men and women. Furthermore, the Ministry of Tourism reorganized and sponsored 'The Festival of Colors' as well as two other ceremonies of the 'Tharu community of Chitwan.'

4.3.2.2. Hybrid Culture

In Pefkochori, Greece, the western and native cultures have coexisted to form a modern society known as the hybrid culture. Borrowing artificial elements and incorporating them into traditional dance, music, events, and activities are examples; serving hybrid food rather than traditional salads and three course meals instead of desserts in the evening show; and opening a bar and disco are other examples (Wickens, 1994). Despite their cultural loss, the locals are content with their good fortune and prosperity. They believe that tourists help them more than locals, so they must satisfy tourists (Herrero et al., 2011).

4.3.2.3. Commodification of Culture

Cultural commodification causes meaning to shift from holy to profane, and from true to unauthentic (Shepherd, 2002). One of the best examples of commodification is found in Greenwood's (1989) case study in Spain, where commodification is exemplified as a ritual that is exchanged for money in an exhibition by modifying or even losing original value to outsiders. In this regard, Cole (2007) summarizes that tourism collects different cultural components and then adds something artificial to make it eye-catching before packaging and selling it to visitors. Shepherd (2002) said that Chinese food in a Chinese restaurant in Beijing with red walls, hanging lanterns, and Chinese scrolls, if served by Chinese waiters, easily guarantees authenticity. However, serving this food in a restaurant in an American suburb shopping mall with mirrored walls and chandeliers and prepared by a Salvadoran immigrant does not guarantee authenticity. This lack of credibility is difficult to refute (Clifford, 1986). Menon (1993) discovered that in Jaisalmer, India, local arts and crafts are not entirely made in a local village with local stone; instead, potters use materials from another territory to make products shiny in order to attract tourists. He went on to say that the vivid photography of women's bodies in decent temples in promotional materials draws visitors to erotic sculpture rather than religious significance. However, he concluded that 'communication growth, poor economic conditions, job opportunity, cost efficiency emphasis of craftsmen, and the economic multiplier effect' in the policymaker's mind all contribute to this harm. Since it is difficult to differentiate between primary and secondary factors in a destination with multiple effects (Shephered, 2002), it is easier to assume that tourism alone cannot alter or preserve culture.

4.3.3. Managing the Impacts of Tourism on Culture

4.3.3.1. Democratic Management

A democratic management group in the destination is essential for empowering community members and managing socio-cultural issues. In Indonesia's Ngadha village, such a group is created by members of each clan to raise their voices against corruption, collect, and distribute funds, and decide on cultural performances, which is now preserving the culture and reaping the benefits of tourism (Cole, 2007). A unique team of experienced people who are unconnected directly to the business process and intent is needed for product development, who will design the location and time of performances (Shephered, 2002). Furthermore, traditional goods should be marketed to both tourists who purchase them as souvenirs and locals.

4.3.3.2. Managing the Process

The process, not just the product, should be prioritized. From 2004 to 2012, The Blue Yonder (2012) ran a project in India where it added story telling for every success and context history telling at pottery sale points, which increased the tourists' intentions to purchase. Furthermore, hiring an interpreter is preferable to directing. For example, according to this tour operator, the interpretation jobs linked locals with the tourism process, which provided authenticity for tourists. On the other hand, it raised locals' consciousness of the importance of preserving arts, folklore, and other forms of cultural heritage. It is also important to reduce reliance on tourism. Through co-creation, co-management, and training, Blue Yonder developed tourism as an alternative source of income. For example, a day plumber might become an oracle for folklore phrases, or a day mason might become a folk artist in the evening, earning good money while also maintaining culture.

4.3.3.3. Strengthening the Local Cultural Value

The importance of enhancing local cultural significance cannot be overstated. Locals in Indonesia agreed to stage dance and music but opposed staging rituals because they did not want to commodify their core value (Cole, 2007). An organized connection between the formal and informal sectors of the tourism industry is beneficial in reducing tourist inconvenience and conflict between stakeholders. From 2010 to 2013, the Travel Foundation's project in Sri Lanka included training for beach boys, small shops, vendors, and taxi drivers, as well as dispute resolution and capacity building. Following these measures, the beach boys became beach owners, the hassle and friction were minimized, and the locals earned more money as a result of the tourists' repeat visits (The Travel Foundation, 2013). Other significant initiatives can be found in McCombes' (2008) case study in Gambia.

4.3.3.4. Considering the Health Impact

Tour operators, hotels, and even the local community must assess health risks and collaborate with international health organizations. Staff, community members, and visitors, according to Wall and Matheison (2006), all have direct and indirect health effects on others. The International HIV/Aids Alliance is successfully promoting recommendations for representatives, guides, community members, and visitors on sex related issues in Caribbean destinations through toolkits to minimize HIV impacts (HIV/Aids Alliance, n.d.). To mitigate the negative effects of gambling, such as a casino, construction should not occur at the beginning of the tourist destination's life cycle (Butler, 1980). For example, based on a case study in Macao and the United Kingdom, McCartney (2005) proposed that this type of tourism attraction be created at the growth or maturity stage rather than at the beginning.

Some may believe that since the tourism sector in developing countries is growing rapidly, there must be a Western impact. Others believe that all communities and traditions are evolving as a result of their inability to resist outside interference. As a result, it can be fair to assume that tourism contributes, but not just in the way that it affects culture (Sharpley, 1994). However, it is important to have a responsible tourism strategy in place to maximize the positive impact and minimize the negative impact. Many tour operators in South Africa are currently following this type of strategy, which includes tracking impacts, joint planning and decision making, addressing local poverty needs, providing advice, engaging, and collaborating with locals at various stages of tourism (Spenceley, 2007). Cultural tourism is more likely to sell to culturally sensitive citizens, and limiting the number of visitors when the situation is precarious might be a smart strategy. Finally, government measures such as establishing a code of conduct and conserving material and non-material heritage, as well as the establishment of a cultural research center for cultural education, training, seminars, exhibits, and performance, could aid in the responsible management of tourism's impacts on culture.

4.4. ECONOMIC IMPACTS OF SPORT TOURISM

The direct spending of visitors associated with major sporting events has the greatest economic impact. Only expenses that would not have happened in the absence of the incident should be counted. This means avoiding spending by visitors who would have visited anyway but planned their trip to coincide with the event. After all such spending has been avoided, the net remaining expenditure can be calculated using the host economy's input-output model.

Over the last two decades, cities, states, and countries have become more competitive in hosting mega sporting events. As a result, governments, and other supporters of major sporting events typically attempt to back up their arguments that the event would improve the economy by commissioning an economic impact statement. Hosting a sporting event has uncovered a variety of benefits in our communities, with some of those benefits including increased community exposure, positive psychic profits, and improved community image. Economic impact in sporting events is characterized as the net change in an economy caused by activity involving the acquisition, operation, development, and use of sport facilities and services (Lieber and Alton, 1983), which generates visitor spending, public spending, job opportunities, and tax revenue. Major sporting events can have a major economic effect on the host town, city, or region. Countries hosting the mega sports event should develop a plan for major investments in sports stadiums and other facilities. It is widely believed that the magnitude of such an occurrence, as well as the timeline for its planning, would result in significant and long-term economic benefits for the host nation. This would result in enormous economic benefits with a substantial economic effect. Hosting a mega sport event like the Olympics has shown a variety of significant and long-term benefits in our communities, such as improved community appearance, increased community awareness, and positive psychic profits, which are all normal and acceptable hypotheses. However, sport is more than just a form of entertainment; it also serves as a full-fledged industry, with the economic effect serving as a foundation for the development of many related industries in communities. The economic outcomes of sporting activities, whether positive or negative, can be a key factor in determining a community's draft budget for the coming year.

The value of this sector's economics can be measured by the number of people employed in it as well as the income produced. A significant number of people are working in the tourism industry, either directly or indirectly. It involves transportation, lodging, food, ticketing, guiding, boating, rafting, trekking, and the upkeep of tourist attractions, among other things. The host offers all necessary services and amenities to the guests. Both of these services are paid for by the guest. As a result, tourism provides significant benefits for both the people of the tourist's origin and the people of the destination. The development of infrastructure in various tourism-related activities allows for improved movement facilities for both incoming and outgoing visitors. One of the primary reasons for developing a country is to encourage tourism destinations for international visitors. It aids in the acquisition of foreign exchange. A sufficient amount of foreign reserves aids in the maintenance of a country's trade balance. Aside from improving the country's trade balance, it also creates job prospects for the country's unemployed youth. This is how they make a living and produce money for their survival and well-being. However, tourism is not always beneficial to the economy; it may also have a negative impact.

4.4.1. Positive Economic Impacts of Tourism

Tourism has many positive impacts on the economy of the local area of the tourist's destination, particularly in less economically developed or developing countries. Among them are:

- Expenses incurred to promote tourism raise revenue in the host countries. This, in turn, boosts growth in other related economic sectors.
- As previously mentioned, attracting international visitors to a host country generates foreign currency. This is a very good indicator of having strong foreign trading capabilities.
- The government of a host country generates revenue through taxes on income from tourism employment, business, and entry fees at various tourist sites/monuments, as well as toll taxes, and so on.

- Many of the products required by tourists are imported from other countries. When the government imposes import duties, it profits financially.
- Rapidly expanding national and international tourism has resulted in numerous job opportunities. It has aided the economic status of those involved in tourism, whether directly or indirectly, through hotels, restaurants, guides, nightclubs, taxis, local crafts, paintings, and cultural items, among other things.
- Tourism encourages the government to spend more in various forms of infrastructure, resulting in better services for both visitors and locals.
- Tourism also promotes local people by providing opportunities for employment in informal sectors such as street vendors, taxi drivers, tea/coffee stalls, magazine corners, shops selling packaged food, and so on.
- Employment in the informal sector contributes to the local economy. It has a rippling impact by saving and creating more revenue for those people at the same time.

4.4.2. Negative Economic Impacts of Tourism

Tourism has many negative impacts on the economies of tourist destinations, especially in less economically developed countries (LEDCs) or developing countries. Among them are:

- There are many hidden costs associated with tourism that have a negative impact on the economies of the destination countries. Most of the time, richer countries reap more than poor countries. Many consumable goods, such as food and beverages, must be imported by host countries that are poor or developing. This is due to the low quality of local goods. As a result, revenues are transferred to developing nations, where they are transported.
- In all-inclusive package trips, almost two-thirds of the cost goes to airlines, hotels, and other foreign firms, rather than to local businesses or staff.
- In all-inclusive package tours, visitors typically spend their entire stay on the same cruise ship or resort, which offers all they want. There are no other options for generating revenue for the local economy.

- The money is used by the government to improve tourism infrastructure in poor and developing countries. However, as foreign investors fund resorts and hotels, the benefit is exported to other developing countries.
- Tourists' increasing demand for basic services and products causes price increases in the destination country. It has a negative impact on local residents whose income does not arise in the same proportion.
- Tourism development in a destination country raises the value of facilities and property. As a result, meeting the basic needs of the people in that area becomes extremely difficult.
- Because of seasonality or harsh climatic conditions, many countries do not promote tourism. Locals can find work during the tourist season, but there is no guarantee that they will find work again the following season. As a result, they are concerned about their ability to earn a living.
- People are relocated in order to build airports, resorts, hotels, nature reserves, historical, and other tourist attractions, and other tourism development projects.

4.5. POLITICAL IMPACTS

Tourism has an impact on politics rather than politics having an impact on tourism. However, the governments of developing countries, from which well-heeled tourists travel, indirectly influence the governments to provide hospitable conditions for visitors. Everyone is worried about the safety and security of the trip to and from the tourist destination. The main goal of tourism is to learn about the region being visited as well as to have fun and enjoy yourself. Tourists are for them, but they want to go to places that will help them achieve their goals. They do not feel safe and comfortable in all tourist destinations around the world. A tourism strategy directs the tourism industry. This policy is a set of laws, regulations, guidelines, and directives for tourism development and promotion. It establishes a structure through which collective and individual decisions affecting tourism development can be made. The government should develop a tourist code of conduct. Some perspectives favor tourism, although others discourage it. They are as in subsections.

4.5.1. Positive Political Impacts

There are numerous positive political impacts on tourism, especially at destination locations. Among them are:

- Political stability is of primary concern to tourists. A significant number of tourists are drawn to any tourist destination by its safety and security.
- Law and order are the priority of the government. It aids in the inflow of visitors if it is well handled.
- Political stability and tourist security from government machinery is a tonic for tourism promotion.
- Political will to build infrastructure in tourist-attraction areas promotes tourism. It makes it easier for visitors to get around, stay in hotels, get directions, and use other amenities.
- Tourist-targeted planning hastens the impact on tourism.
- The beautification of a site and the provision of protection without significant modification to the natural environment often draws a large number of visitors.
- Tourist decision-making about destinations may be influenced by risk perception. It can have a major impact on tourist inflow to a destination.

4.5.2. Negative Political Impacts

There are many negative political impacts on tourism, particularly at the destination places. Important among them are:

- Political instability and conflict at the destination area/country proves to be a hindrance. In fact, political instability is a condition of a country where the government is unstable or is being toppled. Sometimes, the government is run by a faction follower of the coup. In such conditions, law, and order is a big problem. Therefore, tourists refrain from visiting such destinations.
- Sometimes, the governments of the tourist origin countries give a warning or issue advisory to their citizens not to visit such countries where political instability or conflict prevails.
- When the tourists are already visiting the country where some political instability comes in, a warning is issued to the citizens

to leave the country at the earliest.

- The unstable government is not able to maintain law and order situation for the citizens of the country. They may not be able to take proper care of the foreign tourists. Hence, it becomes a deterrent to tourism.
- Government is the only authority to formulate a policy of any concern. All the policy matters concerning the tourism are taken care of by the government, which in turn affects the outcome of the policy.
- A policy adopted in favor of tourism and infrastructure development leads to the growth of tourism and in inverse conditions, the result is contrary.
- Terrorism activities or communal conflicts in any part of a country or globe reduces the chance to develop tourism.

4.6. IMPACT OF SPORT ON ENVIRONMENT

The complete surroundings or conditions in which a human, animal, or plant resides or interacts are referred to as the environment. A person's environment is made up of everything that is directly or indirectly related to their living conditions. It includes homes, buildings, people, animals, plants, soil, water, temperature, light, air, flora, and fauna, and other human-made infrastructures, among other things. Living plants and animals not only live in the environment, but they also communicate with one another. Such interactions have a major effect depending on the dynamism of the interaction. Tourism constantly brings a large number of people with diverse attitudes to a location. The large number of people arriving at certain locations places a significant strain on different services. More infrastructure is being built to accommodate them, causing significant changes in the landscape of the region. Some of the environmental changes have also been found to draw an increasing number of visitors. As a result, tourism contributes to environmental preservation. Now let us see those positive and negative impact on the environment of a destination.

Sport, like any other human activity, is embedded in the physical world and is bound to have an effect on it as well as be influenced by it. The origins of global environmental problems can be found in local environmental conditions, and the relationship of sporting cultures with the world in which their practices are conducted must be examined in this sense.

Sport and the world have a two-way partnership. The impact of the environment on sport is more palpable and immediate, influencing the scheduling of specific sporting activities based on the suitability of the climate and physical environment of the specific location. Global warming clearly has the potential to have a long-term negative impact on sports in general, and winter sports in particular. The emphasis of this segment, however, is on the second aspect of the two-way relationship, namely the impact of sport on the environment.

4.6.1. Third Pillar Added to the Olympic Movement

These impacts may be categorized as short-term, long-term, direct, or indirect on a larger scale. Certain impacts are common in nature and can be associated with most sports, while others are exclusive to specific sports. Recognizing the gravity of the situation, the International Olympic Committee (IOC) has designated 'climate' as the third pillar of the Olympic Movement, alongside 'sport' and 'culture.'

In 1987, the Brundtland Commission described a "sustainable society" as "one that meets the needs of the present without jeopardizing future generations' ability to meet their own needs." This precise concept calls on the 'society' as a whole to synchronize its activities in accordance with the patterns required by the goal of achieving sustainable development. This is the lens through which we will examine sports here.

Measuring the environmental effect of sports in general, and a specific sporting event in particular, is undoubtedly a difficult job. Measuring an event's 'ecological footprint' may be a useful method in this regard. The definition of ecological footprint specifies a hypothetical region of land that would be needed to replenish the total amount of energy and resources consumed during the case. Energy consumption is transformed into the forest cover needed to absorb carbon emissions, while food consumption is converted into agricultural land capable of producing that amount of food.

The overall ecological footprint is the sum of these two factors. As previously said, the impact of sports on the environment can be categorized as short-term, long-term, direct, or indirect, and all of these factors contribute to the overall ecological footprint.

Short-term impacts include noise and air pollution during the event, while long-term impacts are those that persist after the event has ended. For example, biodiversity loss as a result of infrastructure development or environmental degradation in alpine areas as a result of skiing facilities.

There is also a distinction between indirect impacts. Although the burning of fuel by competing motor vehicles in a motor race is a direct impact, when spectators arrive at the venue-burning fuel on the way-it is an indirect impact. Indirect impacts may also occur as a result of the modernization of the host city of a major sporting event, as well as the resulting tourism avenues that disrupt the ecology of that area.

The magnitude and gravity of the effect are primarily determined by the type of sport and the size of the occurrence. Few environmental impacts are shared by all sports. For example, the development of the playing area by changing the natural environment, or the massive amount of energy expended in transporting countless players and spectators from one part of the world to another during the year.

Few other impacts are unique to particular sports. For example, the massive amount of high-quality fuel burned in a variety of motorsports ranging from Formula One to Nascar; the chemical elements used to harden the ice for alpine skiing; or the vast amount of water used to irrigate golf courses around the world. Cooling facilities in stadiums also contribute to the environmental footprint. So much for sports' negative environmental impact. But, given that every human activity has a significant ecological footprint, the question arises as to why, of all environmental issues, we should be concerned with sports at all.

It includes millions of participants and stakeholders from all over the world, and the environmental consequences are global. It cannot be dismissed as merely a recreational practice. Sports, in particular, have the ability to serve as a classic model for the neoliberal pursuit of evoking cooperation and reform through the collective activities of governments, IGOs, NGOs, and the private sector.

4.6.2. Aim to Devise Ways to Make Sport More Sustainable

To that end, the IOC has added sports as the third pillar to the Olympic Movement. Its charter states as an objective, "...to encourage and support a responsible concern for environmental issues, to promote sustainable development in sport and to require that the Olympic Games are held accordingly." It has established its own 'Agenda 21: Sports for Sustainable Development,' which has included International Federations, National Federations, National Olympic Committees, athletes, teams, and even the sporting goods industry within its purview. Its primary goal is to find ways to make sports more sustainable while also encouraging initiatives that raise awareness through conferences, information sharing, training seminars, toolkits, and resource manuals. The tenets of this document are largely idealistic in nature, and there is no mechanism for enforcing them. Furthermore, non-Olympic activities are exempt from its scope.

Another significant feature of modern sports is the growing corporatism that has changed the sports management model. Global sports have come to symbolize the logic of global capitalism. Marketing and advertising are extremely important for the success of a specific event or sport in general. Cutthroat competition is the norm, and it has engulfed all stakeholders, from players, teams, and clubs to sponsors, sports marketing companies, kit suppliers, and advertisers.

This intense competition among all the stakeholders may not leave the scope for participation on any issue-leave alone environment. If there is less interaction, there will inevitably be more creativity.

Thus, we have seen some amazing developments that are important to sustainable growth, and not surprisingly, the majority of them have come from motorsports, the most corporately structured sport of all. Innovative technologies such as the kinetic energy recovery system (KERS) have the ability to completely change the transportation industry. However, much work remains to be done in order to make this and other similar "clean technologies" accessible at reasonable prices so that they can become more widely used.

Another example is the use of recycled materials in the production of sports apparel. Since leather is derived from nature, synthetic products have taken their place. Sports organizers have recently been concerned about minimizing their ecological footprint, and attempts have been made to plan sustainable activities, i.e., events that generate as much energy as they consume through initiatives such as the use of less carbon-intensive technology and the installation of solar panels in stadiums.

Sports provide numerous opportunities to raise environmental awareness, create capacity, and take far-reaching steps for environmental, social, and economic growth in society, and this opportunity must be capitalized on. However, as is always the case, in the absence of legally binding enforcement mechanisms, the "tragedy of the commons" doctrine is bound to stymie progress in this field. Under these conditions, it is very possible that sports organizations express concern for the environment merely to instill a logic of false consciousness in the Marxian meaning of the word. The true motivation for this concern could be to function as a delusion in order to stifle the possibility of demonstrations from NGOs or the wider civil society against the sporting bodies. Grand and unsustainable opening and closing ceremonies, constant construction of new stadiums, and an uncertain plan for the use of these venues demonstrate that promoters are unable to compromise on the spectacle value of sports. As a result, the green light appears to be on as long as technical advances related to sustainability are economically advantageous to stakeholders.

4.7. EVENT IMPACT MANAGEMENT

As mentioned in the previous section, best practices in event management necessitate a constructive approach to positive and negative impact management.

You can reduce the negative effects of your event's preparation and staging by recognizing possible negative impacts and designing methods to remove or at least mitigate these. Similarly, by understanding the possible and desired positive consequences of the event, strategies, and plans can be enforced to achieve the desired results.

4.7.1. Minimizing Environmental Impact

The negative environmental impacts that can occur as a result of event staging are a common source of concern. All activities will have some kind of environmental impact, which will naturally increase in proportion to the event's scale.

Pollution from automobile pollution, noise, chemicals, sewerage, ammonia, and other common negative environmental impacts include: Litter in the event area and on the event ground. Litter and waste products are disposed of in landfills, composting plants, or recycling facilities. Congestion, as well as the resulting local air pollution from car emissions. Greenhouse gases emitted as a result of the use of gasoline or electricity. Environmental degradation-as activities are held outside in vulnerable areas.

Following the principle of reducing, reusing, and recycling is a strong starting point for minimizing the negative impacts of the event. It should be noted that you should begin with and prioritize the reduction of waste outputs and resource inputs, such as reducing all waste outputs and the use of non-renewable energy inputs.

After all conservation efforts have been completed, the next step is to investigate where resources can be reused. For example, reusable cups and other catering products are still preferable to recyclable ones. Similarly, if products such as promotional banners can be reused in the future, there will be less waste and lower prices. It should be remembered that, while preferable, recycling is the last resort before landfill. This means that the focus should always be on reducing and reusing items, and recycling should be enforced only when non-reusable materials are used. Recycling stations at events should be easy, accessible, and reviewed on a regular basis.

4.7.2. Maximizing Tourism and Economic Benefits

Events should always aim to produce economic benefits for the host community and the surrounding region. This is not only good ethical practice, but depending on the magnitude of the impacts, it can also free up potential financial tools and stakeholder support that would not otherwise be available. Economic benefits are often derived from tourist's spending money in the region.

Economic benefits are frequently derived from visitor's spending money in the region; therefore, any event that attracts visitors into an area would provide an economic benefit. To maximize the profit, tourists should be encouraged to stay in the area and use local services/businesses. As a result, activities can be used to draw tourists outside of typical tourist seasons, promoting the use of underutilized tourist facilities during low-demand periods.

Economic effects may be direct, resulting from the event's staging, or indirect, such as providing a favorable experience for event participants, which increases the probability of repeat visits and positive recommendations both online and offline. As previously stated, the degree of local sourcing will also relate to the economic benefit gained from your event's hosting; indeed, local sourcing of food and drink is one of the goals of the Welsh Government's Local Sourcing Action Plan, for example. Using local suppliers wherever possible would almost always result in a higher return for the host community than bringing in suppliers from further afield. Using local suppliers can also help to create a "sense of place." Sense of place includes, in addition to food and drink, the following:

- ***People:*** Their abilities and expertise, special interests, and personal stories.
- ***Local Products:*** Arts and crafts.
- ***Heritage and Culture:*** Local history, traditions, language, folklore.

- ***Community Activities:*** Agricultural shows, food festivals, farmers markets.
- ***Environment:*** Natural and built landscapes, landmarks, flora, and fauna.

4.7.3. Minimizing Negative Social Impacts

Social impacts are topics that affect the community and should be carefully addressed during event planning and execution. Negative neighborhood impacts may include problems such as traffic, pollution, and a lack of access to community services and amenities, which can contribute to feelings of dissatisfaction and isolation from the community. Events may also lead to crime and antisocial behavior if prevention techniques to address problems such as alcohol and drug misuse are not enforced.

Maximizing social benefits necessitates some level of community consultation and participation in the planning process, which can be accomplished by community forums, questionnaires, focus groups, and other means. Reinvestment and funding for local organizations and charities will also help compensate communities for the temporary disruption caused by an event.

Chapter 5

Sport Tourism and Rural Development

CONTENTS

5.1. INTRODUCTION

Tourism is one of the world's biggest and most lucrative sectors. It has a major societal impact, such as job creation, economic growth, and infrastructure development. It is possible to raise the income level of local people who make their living primarily from agriculture with the aid of rural growth, as well as to improve social life conditions. Rural life can be combined with the manufacturing and service sectors to reduce unemployment. There must be certain prerequisites for tourism activities to occur in any area. There is a need for infrastructure and superstructure investments, which are essential for rural tourist attractions. The countryside's attractions are events that seek to maximize the benefits of nature. Some of these activities include walking, hiking, fishing, cycling, and ecotourism. Cultural tourism and mass tourism activities will provide rural tourism activities together with or complementary to rural development. After realizing the potential economic impact of this industry, the government has developed a number of policies to encourage it in recent years. In this chapter, the relationship of sport tourism activities in rural development is used to examine rural development and tourism styles. In rural areas, the positive and negative aspects of tourism activities are investigated.

5.2. RURAL TOURISM: A SUSTAINABLE APPROACH TO RURAL DEVELOPMENT

Rural tourism is described as tourism that allows visitors to experience "the actual culture" of a rural environment. Tourism can play a significant role in diversifying the rural economy, and its expansion in rural areas contributes to the long-term viability of the population and economy of these areas, as well as providing the foundation for achieving long-term rural growth. As a result, the aim of this chapter is to look into the relationship between tourism and rural development. The pastoral life of rural areas is an alien idea to the tourist who lives a westernized lifestyle. As a result, the rustic beauty becomes a tourist attraction (Figure 5.1).

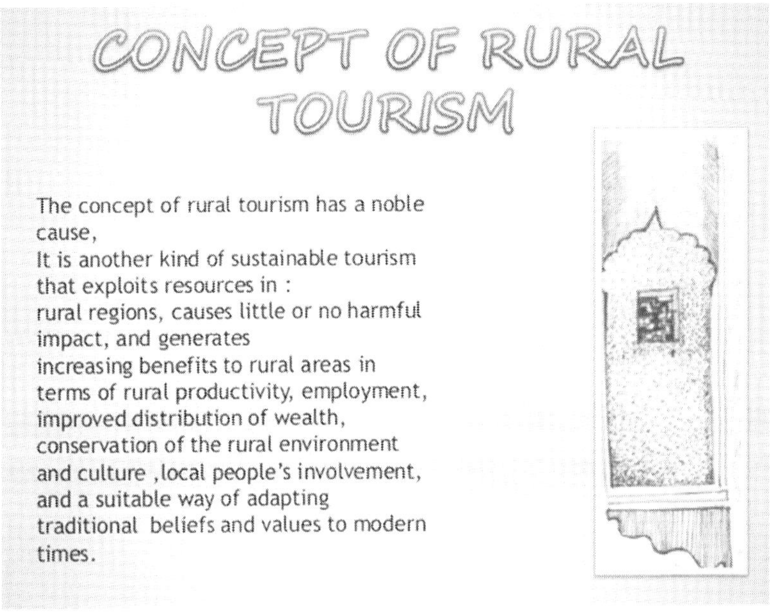

Figure 5.1. Rural tourism is described as tourism that allows visitors to experience "the actual culture" of a rural environment.

Source: https://image.slidesharecdn.com/rural-161129190712/95/rural-tourism-9-638.jpg?cb=1480446579.

5.2.1. Defining Rural Tourism

Rural tourism is characterized as the movement of people from their usual place of residence to rural areas for a minimum of 24 hours and a maximum of six months for leisure and pleasure. Rural tourism encompasses all tourism events that take place in rural areas. The term "rural tourism" is not well known and can be interpreted in a variety of ways.

Fleischer and Pizam equate rural tourism with the "country vacation," in which the visitor spends the majority of his or her vacation time engaged in leisure activities on a farm, ranch, country home, or in the surrounding areas.

According to the OECD, a rural area is described as "at the local level, a population density of 150 persons per square kilometer is the preferred criterion." At the regional level, geographic units are classified into three categories based on the proportion of their population that is rural: predominantly rural (50%), significantly rural (15–50%), and predominantly urbanized regions (15%).

The Council of Europe implemented the term 'rural area' to denote the following characteristics.

Its characteristics include portraying native arts and heritage in their natural environment (rural areas), as well as experiencing local culture through direct interaction with village residents. Rural tourism, cultural tourism, adventure tourism, nature tourism, and eco-tourism are only a few of the many facets of rural tourism. Unlike traditional tourism, it does not include monuments or distinct cultural structures that impart information about the area.

Thus, rural tourism, with its diverse offerings of traditional arts and culture, provides visitors with a respite from their hectic, fast-paced lives enslaved by new technology and urban ways of life. It is a "vacation" that provides the visitor with a one-of-a-kind learning experience.

A stretch of inland or coastal countryside, including small towns and villages, where the majority of the land is dedicated to:

- Agriculture, forestry, aquaculture, and fisheries;
- Economic and cultural initiatives of country-dwellers;
- Outdoor recreation and recreational areas outside of cities, such as nature preserves;
- Other uses, such as housing.

According to Dernoi, rural tourism might be conceived as tourism activities in a 'non-urban territory where human (land-related economic) activity is going on, primarily agriculture: a permanent human presence seems a qualifying requirement.'

The OECD states rural tourism should be:

- Situated in rural areas.
- Functionally rural, based on the unique characteristics of the rural environment; small-scale industries, open space, interaction with nature and the natural world, history, traditional cultures, and traditional traditions.
- Rural in scale-both in terms of construction and settlements- and therefore limited in scale;

- Traditional in character, rising slowly and organically, and linked with local families;
- Sustainable-in the sense that its construction can contribute to the preservation of an area's unique rural character, as well as in terms of resource usage.
- Many different kinds, each reflecting a different aspect of the rural climate, economy, and history.

5.2.2. Types and Forms of Rural Tourism

Rural tourism refers to any form of tourism that highlights rural life, art, culture, and heritage in rural areas, thereby benefiting the local community economically and socially and allowing interaction between tourists and locals for a more enriching tourism experience.

A variety of words are used to describe tourism activity in rural areas, including agrotourism, farm tourism, rural tourism, soft tourism, alternative tourism, eco-tourism, and others, the definition of which varies from country to country.

The European Community (EC) has adopted the word "rural tourism" to refer to all tourism activities in a rural community.

The following are examples of rural tourism types and forms:

1. ***Agrotourism:*** Although the term "agrotourism" is frequently used to describe all tourism activities in rural areas, it more frequently refers to tourism products that are "directly connected with the agrarian environment, agrarian products, or agrarian stays": staying at a farm, whether in rooms or camping, educational visits, meals, recreational activities, and the sale of farm products or handicrafts.
2. ***Farm Tourism:*** Explicitly farm-related tourism involves staying in farm accommodations and finding experiences from farm operations and attractions.
3. ***Wilderness and Forest Tourism:*** This form of tourism involves tourists exploring the wilderness and natural beauty of the rural region. It may be indirectly included within tales of rural tourism, or it may be considered distinct. Tourists who participate in wilderness and forest tourism visit the natural habitats of plants and animals.

It primarily refers to non-consumptive encounters with wildlife and nature, such as watching and photographing animals in their natural environments. Wilderness and forest tourism encompasses a wide range of tourism activities such as wildlife photography, safari, and bird watching, trekking, and hiking, among others.

4. ***Green Tourism:*** It refers to tourism in the countryside or in green areas. It is most widely used to describe environmentally sustainable types of tourism as opposed to conventional, mass tourism. Green tourism is a significant aspect of rural tourism in rural areas.

   Green tourism is portrayed as a tourism creation strategy that aims to cultivate a symbiotic relationship with the physical and social ecosystem in which it relies, whilst still implicitly aiming for sustainability ideals.

5. ***Ecotourism:*** It is a form of nature tourism (tourism to natural, unspoiled areas) that includes the active promotion of environmental protection and direct benefits for local communities and cultures, as well as the provision of a meaningful, educational experience for tourists. Ecotourism is a category of sustainable tourism activities that take place in the natural environment.

5.2.3. Rural Tourism Activities

According to various figures, rural tourism accounts for 10 to 20% of total tourism production, and a Eurobarometer survey found that 23% of European holidaymakers select the countryside or rural areas as a destination each year.

Rural areas host a variety of cultural and natural events. The rural ecosystem has long been controlled for leisure purposes, and this symbiotic relationship has had significant effects on both the environment and activities.

The countryside offers a variety of tourism and leisure opportunities. These behaviors can be classified based on their nature. The following are examples of rural tourism activities:

- Cultural events;
- Touring;
- Water-related activities;
- Passive activities;
- Aerial activities;

- Sporting activities;
- Hallmark events;
- Business-related activities.

Hiking, horseback riding, traveling in gypsy caravans, motorized touring, small village/town touring, walking, adventure vacations, and wilderness vacations are all examples of tour activities.

Rural areas offer a variety of culturally based tourism activities. Archaeology, restoration sites, rural heritage studies, museums, craft courses, and creative expression workshops are some examples of rural tourism cultural activities.

Water-related activities in rural tourism include fishing, swimming, river tourism, canoeing, kayaking, windsurfing, speedboat racing, and sailing.

Rural areas have a wide range of rural tourism sporting events. Potholing, rock climbing, orienteering, tennis, golf, low-intensity downhill skiing, and hunting are a few examples.

Rural tourism activities that are common include fitness training, assault courses, spas, and health resorts. A substantial number of tourists travel to rural areas solely for the purpose of improving their health through health-related tourism activities.

Passive activities such as relaxation holidays in rural settings, nature research in outdoor settings such as bird watching and photography, and landscape appreciation are also well-known rural tourism activities.

5.2.4. Characteristics of Rural Tourism

The idea of rural tourism is for a good cause. It is a form of sustainable tourism that exploits resources in rural areas, has little to no negative impact, and provides increasing benefits to rural areas in terms of rural productivity, jobs, improved wealth distribution, conservation of the rural environment and culture, local people's involvement, and a suitable way of adapting traditional beliefs and values to modern times.

The following are some characteristics of rural tourism:

- Seasonality;
- Fragmentation;
- External market needed;
- Co-operation needed between internal and external market;
- Role of women;

- Economic role: side income for farmers as well as other entrepreneurs throughout the rural area.

Rural tourism will help to boost local performing arts, preserve local culture, and prevent rural migration. Rural tourism has the potential to attract visitors by offering an excellent glimpse of village ambiance with local cuisine.

5.2.5. Benefits and Drawbacks of Rural Tourism

The advantages and disadvantages of rural tourism can be categorized according to the stakeholders as follows:

1. Benefits for Residents:

i. *Rapid and Holistic Infrastructure Development:* A tourist destination must be appealing to potential visitors; it must have basic amenities such as restrooms, electricity, and water supply, sanitation, and sewerage facilities. These areas' accessibility would also need to be improved. The growth of rural areas as tourist destinations would result in the comprehensive development of their infrastructure. The allure of economic benefits will push authorities to grow these areas more quickly.

ii. *Connectivity with the Outside World:* Rural areas are currently considered isolated areas with little access and interaction — people communicate with tourists from neighboring regions. With the advent of rural tourism, people from a variety of geographical and educational backgrounds can visit these underdeveloped areas and engage with the locals, thus expanding their knowledge and skills.

iii. *Skill Development and Job Creation:* As the idea gains traction, a significant number of visitors would flock to rural areas, increasing appreciation for local arts and crafts. In addition, residents will need to be educated in skills related to the hospitality and tourism industries; new work profiles such as guides, event planners, accommodation managers, caterers, and so on will arise.

iv. *Improved Rural Economy:* As visitors visit these areas, the economic situation of the rural areas will improve. With the rise of new hotels / B&Bs, improved sale of local arts and crafts, and prerequisite of means of transport, the income of the resident population will undertake immense growth.

v. *Improved Quality of Life:* The cumulative impact of the individual benefits would result in a significantly improved quality of life for residents.

2. For Tourists:

i. *Experience a Different Lifestyle:* Tourists who spend the majority of their time in cities will have the opportunity to escape the frantic environment of cities and experience the tranquil atmosphere of villages.

ii. *Get Closer to Nature:* Over time, urban areas have moved away from nature, while rural areas develop in symbiosis with nature.

3. The Challenge:

i. *Adapting to a Pastoral Way of Life:* How simple will it be for tourists to adapt to a pastoral way of life in villages? The balance between preserving village infrastructure and improving facilities for tourist convenience must be considered.

ii. *Increased Demand for Natural Resources:* As more visitors visit these regions, they can consume more food and water, raising the demand for these resources.

iii. *Waste Management:* As the visiting population grows, so will the amount of waste have produced. Waste production and its effects, such as plastic waste, must be regulated.

iv. *Adaptability of Native Population:* The resident population, which is not averse to urban culture, especially of people from other countries, may have trouble accepting their new visitors. To avoid disrupting the native population's community, the authorities would have to interfere by legislation.

v. *Environmental Quality:* As waste production and pressure on natural resources increase, so will environmental quality. Maintaining harmony with the ecosystem would be a significant challenge; failure to do so may be fatal to the endemic flora and fauna.

Rural tourism is thus a developing phenomenon that is attracting the attention of government, organizations, and individuals. Schemes like these are not only an appealing investment, but also a long-term step toward rural growth. Rural tourism, as a viable alternative to prevalent environmentally harmful activities, maybe one of the most beneficial types of tourism, benefiting society, people, tourists, and the environment!

The current administration wants to raise the living standards of "farmers." It cannot stop the spread of urban sprawl, nor can it exert control over agricultural products in the event of bad weather. Rural tourism has the potential to save the rural economy. As a result, the potential of rural tourism must be fully explored and exploited. Authorities must capitalize on the idea, which generates income while also signaling economic growth and social advancement.

5.2.6. Promoting Rural Tourism

Tourism has been thriving and expanding since the dawn of time. This smokeless industry has received special attention in recent decades. Tourism has developed as a significant and well-organized sector, with benefits distributed to broad segments of the population. Kashmir Valley, also known as the "Paradise on Earth," is one of the regions of India that offers a range of experiences to tourists. By generating income in the tourism sector, we can achieve the required stability, well-being, and high quality of life for the citizens of the area. Tourism's growth capacity can be seen as a rural development strategy. The development of a strong forum centered on the idea of rural tourism is unquestionably beneficial for a country like India, where nearly 70% of the population lives in villages. There is a need to learn about the role of rural tourism in the region's long-term growth and to improve rural tourism development. The development of rural tourism and the improvement of society are inextricably linked.

Rural tourism can be described as any type of tourism that showcases rural life, art, culture, and heritage in rural areas, thereby benefiting the local community economically and socially and allowing interaction between tourists and locals for a more enriching tourism experience. Rural tourism is an operation that occurs in the countryside. It may include farm/agricultural tourism, cultural tourism, nature tourism, adventure tourism, and eco-tourism. Unlike traditional tourism, rural tourism is characterized by the following characteristics: it is experience focused, the locations are sparsely populated, it is primarily in natural environments, it integrates seasonality and local activities, and it is centered on the preservation of culture, heritage, and traditions.

The rural tourism industry in India's Kashmir Valley has the potential to develop. Tourists tend to visit new locations that are close to nature and away from the hustle and bustle of the region. The essence of rural areas in Kashmir Valley is not polluted to the same extent as in other states in India, so there is enough potential for the development of rural spaces for the development of tourism in the region. The tourism sector is also known as the "Smokeless Industry" because it has more benefits than drawbacks due to the lack of a chimney. Only by developing rural tourism spaces can sustainable tourism growth be possible. Tourism is the most important part of the tertiary market. In reality, one out of every 16 jobs in the world are employed in the tourism industry. By the turn of the century, tourism will be the world's leading export sector. As a result, the creation of rural areas could not only provide us with a clean climate but also generate revenue in the form of tourism. Thousands of rural people are drawn to this industry and find work in it. Rural tourism is frequently regarded as intrinsically sustainable since it draws a limited number of visitors, there is no need for substantial infrastructure growth, and tourists are typically genuinely interested in the local culture and traditions. One of the keys draws of rural vacations is the personal contact with local people, which allows hosts and guests to exchange ideas and knowledge, and thus tourism can fulfill its position as the "industry of harmony," as a medium for mutual understanding.

However, when the growth of rural tourism is examined in greater depth, some concerns about intrinsic sustainability emerge. The most important problem to examine is the economic viability of rural tourist services since demand is often seasonal, occupancy rates are poor, and the expenditure needed to create or improve tourist facilities is often large. Tourism is unlikely to be a viable sole source of income in most rural tourist destinations. Rural tourism is typically only one of several income-generating options in an area, so its position in long-term growth is heavily influenced by the performance of other economic sectors, especially agriculture. The Valley of Kashmir is located at the top of India's globe. Kashmir is part of the state of Jammu and Kashmir, which also includes the regions of Jammu, Ladakh, and Gilgit.

The physical places, the compatible seasonal division, the richness of flora and luscious fruits and enticing vegetables have captivated everyone's imagination in the Valley of Kashmir, and as a result, it is rightly referred to as "Paradise on Earth." Nature has endowed Kashmir Valley with so many location advantages that it has the potential to become a tourist hotspot. It is surrounded by beautiful mountains and snowcapped peaks, making your stay in Srinagar enjoyable. Excellent scenic, recreational, religious, and cultural sites are just a few kilometers to 40 kilometers away. Even driving, walking, or riding a horse to these locations is enchanting. The Valley's main tourist attractions include snowy peaks, frozen lakes, ski slopes, pine forests, and freshwater streams.

Tourism's growth capacity can be seen as a rural development strategy. The development of a strong forum centered on the idea of rural tourism is unquestionably beneficial for a country like India, where nearly 70% of the population lives in its 7 million villages. Industrialization and growth movements around the world have taken an urban-centric approach. In addition, the pressures of urban living have resulted in a "counter-urbanization" syndrome. This has increased interest in rural areas. Rural tourism is one of the few things that can help to solve many urban problems. In addition, there are other factors that are changing the trend toward rural tourism, such as increased awareness, increased interest in heritage and culture, enhanced accessibility, and environmental consciousness. This has resulted in a new form of tourism in developing countries, with visitors visiting village settings to experience and live a comfortable and safe lifestyle. This idea has evolved into a formalized type of rural tourism. As a result, we must prepare the same idea in our country as a whole, and in the Kashmir Valley in particular.

Rural tourism is a difficult service within rural communities' landscapes and cultural heritage. Tourism has many potential benefits for rural communities. This activity generates local income, jobs, and welfare development, and it is an important contributor to the rural economy. Rural tourism can be a significant source of tax revenue for local governments, as well as a means of promoting the use and selling of local food items. Rural tourism keeps viable traditional jobs from being replaced, provides business opportunities for rural residents, and creates new job opportunities. Rural tourism helps to maintain rural traditions and heritage. It encourages environmental improvements within settlements and acknowledges natural diversity.

5.2.7. Positive Impacts of Rural Tourism

Rural tourism can have a wide range of positive economic, social, and environmental impacts if it is well managed and adheres to sustainable tourism principles. Some of the most frequently mentioned advantages of rural tourism are as follows:

1. ***Creation of New Jobs:*** The creation of jobs is a common positive economic impact of tourism. Rural tourism has the potential to generate a large number of jobs in places where there would

otherwise be few. These employments may be directly linked to rural tourism, such as hotel staff or taxi drivers.

They may also be employed in the rural tourism industry, such as builders (who construct the hotels) or workers employed to maintain and clean the area.

More people working means more opportunities for broader economic benefits. This is due to the fact that employers would almost certainly have to pay taxes on their earnings.

2. ***Benefits to Wider Economy Through Taxes:*** Each destination has its own taxation system. But one thing we can be pretty confident of is that some revenue would be made from taxes on tourism goods and services.

Tax revenue would then be re-invested in other sectors, such as healthcare or education. As a result, tourism has the potential to have a far-reaching positive economic impact (Figure 5.2).

POSITIVE ECONOMIC IMPACT

- It will create employment for the rural people and generate income for them. The villagers will able to provide better food and education for their children.
- Income level will rise.
- Generate foreign exchange.
- Demand for other goods and services will increase.
- Improvement in the public services.
- Generate revenue for the government.
- Modernization of agriculture and other rural activities.
- Local small businessman will be benefited.

Figure 5.2. Positive economic impact of rural tourism.

Source: https://th.bing.com/th/id/OIP.9g6DR3SQbzbHjSbZYxie8wHaFj?pid=ImgDet&rs=1.

3. ***Boosts Local Businesses:*** Locals can set up and run businesses thanks to rural tourism. Rural areas frequently have less well-known chains and brands (think Costa Coffee, Hilton Hotel, etc.), and more small businesses.

   Businesses that are owned and run locally are great because they enable a large portion of the income generated by tourism to remain in the community and avoid economic leakage in tourism.

The local community should take advantage of newly built facilities and services. Rural tourism always necessitates the construction of new infrastructure and facilities.

This is particularly prevalent in transportation networks. Rural areas are inherently underserved by public transportation. Since roads are often narrow and winding, traffic congestion is common, particularly during peak hours.

Rural tourism often results in the creation of new transportation networks and infrastructure, as well as other public facilities and services. This benefits not just the visitors who visit here, but also the local community.

4. ***Cultural Exchange:*** Cultural tourism and cultural exchange are encouraged by rural tourism. Many citizens from various places will move to rural areas for tourism purposes. This allows locals and visitors to get to know one another and learn more about each other's cultures.
5. ***Traditions, Rituals, and Crafts are being Revitalized:*** Tourism has many positive social effects. One effect is that people visiting rural areas are encouraged to share their beliefs and customs with them. This promotes the revitalization and preservation of customs, practices, and crafts.
6. ***Environmental Protection and Conservation:*** Since rural tourism is also dependent on the environment that is visited, there are also programs in place to preserve and restore areas. This involves, for example, designating an area as a national park or naming it an area of outstanding natural beauty. It also involves putting management mechanisms in place, such as limiting visitor numbers or excluding those places from being used.

5.2.8. Negative Impacts of Rural Tourism

Although rural tourism has many benefits, it also has some drawbacks that must be considered. Here are a few examples of the most common:

1. ***Public Services are Under Pressure:*** Tourism is frequently seasonal, with peaks and valleys. In the United Kingdom, for example, rural areas are busier on weekends than on weekdays, and there are more visitors during the school holidays than during the school year.

This can put a lot of strain on public services. During the summer months, when hotel occupancy rates are at their peak, hospitals may be overburdened. On bank holiday weekends, roads can be congested as city dwellers escape to the countryside for some fresh air.

2. ***Increased Price of Land and Real Estate:*** Tourism may cause a rise in land and housing prices. This can have a negative impact on the local community. Some residents may feel compelled to move because they can no longer afford to live in the neighborhood, a phenomenon known as gentrification.

   Others may have a poorer quality of life (for example, a smaller home and less discretionary income) than they would have had if tourism did not exist.

3. ***Congestion and Overcrowding:*** Overcrowding and congestion can occur in rural tourism. This is particularly noticeable during peak periods such as Christmas, the summer holidays, and weekends.

4. ***Inappropriate or Too Much Development:*** Another issue with rural tourism is the possibility of overdevelopment. This can have an effect on a destination's attractiveness to both visitors and locals.

   Any building may not be in line with the area's traditions. A new theme park, for example, will almost certainly change the region (because they are mostly located in rural areas). It will carry a new kind of tourist as well as the related innovations (hotels, food outlets, etc.).

5. ***Rural Tourism Management Techniques:*** It is important to use effective management strategies to optimize the positive impacts of rural tourism while minimizing the negative impacts.

6. ***Limiting Access:*** Unfortunately, several rural tourism areas are inaccessible to the general public. Enabling broad access is a critical component of ensuring equitable and sustainable tourism.

   According to the Equality Act of 2010, 'Tourism providers should treat everyone using their products, facilities, or services equally, regardless of their age, gender, ethnicity, sexual orientation, disability, gender reassignment, religion, or belief, and refrain from making assumptions about the characteristics of individuals.'

5.3. RURAL TOURISM DEVELOPMENT

Rural tourism development is more than just a well-thought-out strategy. Using an actor-oriented perspective, it can be viewed as a complex, ongoing socially constructed and negotiated mechanism involving multiple social actors who constantly reshape and transform it to suit their perceptions, desires, beliefs, and agendas (Figure 5.3).

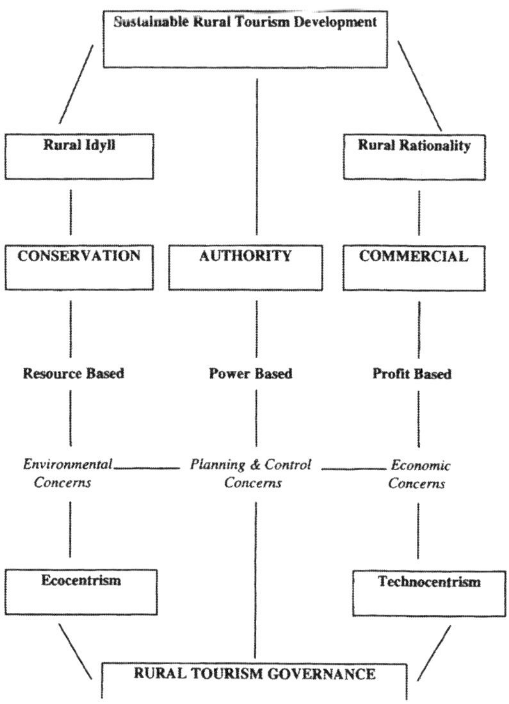

Figure 5.3. A model of sustainable rural tourism development.

Rural tourism development grew in popularity in the 1990s, and a growing body of literature has contributed to our understanding of it as a changing phenomenon.

According to Long and Lane, rural tourism has entered its second stage of growth, with the first being marked by increased participation, product, and business development, and partnership. Long and Lane argue in their study of rural tourism development that rural tourism is entering a more dynamic process of expansion, differentiation, restructuring, and comprehension, at least in Europe and North America. Its second is expected to be more complicated, and it is likely to be so, given the remaining questions about its position in strategy, integration in practice, and dynamic role within the restructuring countryside and broader tourism growth processes. The need for sustainable forms of development is recognized as tourism continues to be grown in rural areas to offset the economic downturn in the primary production sectors. Since the early 1990s, the idea of sustainable tourism development has almost universally been accepted as a beneficial and acceptable solution to, and objective of, tourism development.

The need for economic growth and diversification has guided rural tourism creation, and it is critical that tourism's potential contribution to rural economic growth is not diminished by the over-specificity of sustainable tourism concepts.

5.3.1. Tourism and Rural Development Interaction

Tourism is very important to many emerging economies, despite the fact that they are expanding rapidly. It has been recognized as a sectoral advantage in reducing poverty and promoting growth in emerging and developed economies. Tourism will help local and rural economic growth in the following ways (Ashley, 2000; International Labor Organization, 2011; OECD, 1994):

- Agricultural income has an effect either directly (as an additional revenue source) or indirectly (as a cash flow in the region);
- Offers alternative and complementary solutions for urban and rural life;
- Ensures the continuity of traditional activities (handicrafts, local flavors, etc.);
- Encourages the establishment and development of new investments;
- Ensures the sustainability of rural businesses;
- Establishes an important market for many sectors, providing economic diversification;
- Contains reward applications for local residents;
- Provides funds for sustainable use of natural and historical resources;
- Active use of female labor force in rural areas. Similarly, the negative effects of tourism, especially on the poor, are as follows:
    - More competition for water, land, and other natural resources;

- Rural demand may cause food prices to rise;
- Social and cultural deterioration may occur.

Natural resources are depleted over time as a result of unplanned tourism activities. As a result, the events must be carefully prepared. At this point, the key rule is sustainability. Many research on the relationship between tourism and rural development indicate that tourism provides economic benefits to rural communities. Tourism will benefit rural development when it is well integrated with all stakeholders. The active participation of local people in tourism activities helps to deter immigration by putting the young population to work, and the rise in a variety of agricultural products benefits rural growth. Sustainable use of natural resources and long-term use of the countryside, such as transportation capability and resources, should be prioritized in the preparation of these operations. Accommodation zones, agricultural areas to be opened for tourism, transportation options, and noise and garbage management are all studied in rural tourism infrastructure planning studies. However, if these tourism activities in rural areas are not carried out in accordance with a specific plan and policy, the region will suffer.

5.4. WHAT IMPACT DOES SPORTS EVENTS HAVE ON RURAL COMMUNITIES?

Events, especially those attracting visitors, have the potential to attract outside investment and financial capital into local communities, resulting in positive economic benefits that may sustain rural communities in economic transition. However, there is a tendency to view tourism as a panacea for rural areas' macroeconomic and social problems. However, the social ramifications of tourism and activities are all too often ignored in favor of the economic benefits.

At the community level, social effects may include the development of recreational services and facilities, improvements to infrastructure, behavioral problems, a sense of occasion, enhanced community spirit, enhanced community pride, and a modernized community image. Infrastructure improvements, such as highways, sewage, irrigation, and water supply, while intended to enhance the tourist product, have widespread benefits for host communities as well.

According to Bell Planning Associates (1994), another advantage of tourism growth for local residents is the expanding variety of facilities and services. Reduced access to services and facilities, on the other hand, can have an impact on residents' quality of life, especially during peak periods such as event hosting (Crandall, 1994; Goodwin, 2000). Furthermore, increased litter, pollution, and crowding, noise, and crime have been cited as indicators of strain on local infrastructure and residents' quality of life. Increased sexual harassment, alcoholism, violence, and drug abuse are all significant social issues that can be brought about by tourism growth. Events, in particular, provide the local population with the ability to engage in entertainment, socialization, and a celebratory environment, which is often lacking in rural areas; as a result, the local residents benefit greatly (Janiskee, 1994; Smith and Jenner, 1998). Events have the potential to put small communities, especially rural ones, on the map, contributing to increased tourism potential in the future. Tourism and events have the ability to reinforce the social fabric of the city by growing local engagement and involvement in events, as well as the subsequent socialization.

According to researchers (Janiskee, 1991; Mayfield and Crompton, 1995; Mount and Niro, 1995), an incident in a rural area can be viewed as an act of friendliness and an affirmation of the community's willingness to work together. This is because many activities in rural areas are created by the community and depend on strong community participation and funding, thus developing and leveraging established stocks of social capital. Social capital is the willingness of citizens to safeguard benefits by virtue of membership in social networks or other social systems (Portes, 1998, 2000).

Social capital develops from the social experiences and networks that people encounter on a daily basis. It is not the quality of individuals, but rather the overall quality of a group or culture. Social capital emerges from and contributes to the development of a sense of social trust, the reciprocity standard that underpins social exchange and networks. Cooperation is impossible without confidence, and society-building is impossible without cooperation. Social capital is manifested in public participation norms and networks (Putnam, 1993; Livermore and Midgley, 1998).

Social networks and social institutions are social mechanisms that facilitate interpersonal contact, cultivate strong norms of generalized reciprocity, and promote the development of social trust (Cox, 1995; Sirianni and Friedland, 1995). Social capital is created by active relationships with one another, resulting in the accumulation of trust, and it is increased through continued use (Cox, 1995b). When these stocks of social capital are used, they appear to accumulate, but they may also be exhausted, creating the likelihood of both virtuous and vicious loops that manifest themselves in both highly civic and uncivil societies (Sirianni and Friedland, 1995).

Developing social capital is also a method of increasing human capital by improving individuals' and families' abilities to conquer adversity and capitalize on opportunities (Sanoff, 2000). As a result, rural communities should look to events, especially those created by the community, as a source of community celebration and a source of social capital building within their community through stakeholder community involvement. Individual social effects are related to quality of life, job opportunities, learning new skills, engagement, a personal enjoyment from socializing and recreation with others, and behavioral problems. However, the possibility of full-time continuous jobs in rural tourism and event enterprises is often exaggerated, with a higher proportion of workers previously working as farm laborers, family members, or community volunteers (Hjalager, 1996).

Involvement and participation in activities may also improve an individual's attitudes toward his or her culture (Gorney and Busser, 1996). Participation in event planning and organization also contributes to the acquisition of new skills that individuals can apply to various areas of their lives and those of their group (Reid, 2002a). The ability to socialize and recreate with others is a big motivator, which is especially important in rural communities as a way to forget the harsh reality that they might otherwise face due to drought and distance. While the social consequences of tourism are closely similar to the social consequences of activities, they vary in those events are founded within a community and celebrate the unique characteristics of that community (Delamere, 1997).

Furthermore, the high concentration of event and tourism activity in a small concentrated region over a short period of time produces special and extreme social implications for events. Much of the literature on the social effects of events has concentrated on mega-events or landmark events (Hall, 1992; Hall and Hodges, 1996; Soutar and McLeod, 1996; Crouch and Ritchie, 1999; Fredline and Faulkner, 2000), with little attention paid to the consequences in rural areas (Delamere, 1997; Molloy, 2002). As a result, it is important to examine the social implications of events in rural areas.

Chapter 6

Sport Tourism and Urban Regeneration

CONTENTS

6.1. INTRODUCTION

The issue of regeneration has been a major theme for urban policy and planning in much of the developed world over the last three decades. The effects of seemingly ongoing economic transformation, globalization, and technological and policy change have resulted in structural changes in the foundations of many urban economies, with some areas experiencing high levels of social exclusion and deprivation. As a result of the changes in the urban economy, policymakers have had to reconsider how jobs can be generated and some of the effects of restructuring mitigated. All of this is done in the guise of regeneration.

6.2. WHAT IS URBAN REGENERATION?

Urban regeneration is a comprehensive and coordinated vision and action that aims to solve city problems and bring about long-term improvements in an area's economic, physical, social, and environmental factors. The terms "urban regeneration" and "urban revitalization" are often used interchangeably in the literature. As an area of public policy, urban renewal applies to the redevelopment of economic activity, the restoration of environmental quality, and the recovery of social inclusion.

Rather than planning new urbanization, urban renewal is a method of reorganizing and upgrading existing built environments. It is an ancient idea that is grown over time. Its origins can be traced back to the 1970s, when many cities in the United Kingdom and the United States began programs known as "urban renewal" or "area improvement," which centered on the physical renewal of inner cities defined as "areas of social deprivation" (Figure 6.1).

By the late 1970s, economic considerations such as revitalization of downtown cores or whole neighborhoods had been integrated into renewal plans, and urban regeneration had evolved into a more systematic term. Property-led urban redevelopment schemes dominated urban policymaking in British and US cities in the 1980s, with the belief that a supply of new office, manufacturing, and retail premises would stimulate local economic transformation. These initiatives were part of the strategies for achieving the "entrepreneurial city," a modern type of urban governance designed to promote local economic development.

Figure 6.1. Urban tourism and resilience.

Source: https://moderndiplomacy.eu/wp-content/uploads/2018/10/asean-tourism-asia.jpg.

6.2.1. Why Urban Regeneration Is Very Important Today?

Urban regeneration has many aspects, ranging from renewing the urban environment and restructuring its architecture to regenerating society and developing an overall national identity concept. This type of regeneration could be precisely what a society requires to turn it around, to build new centers and homes.

Urban regeneration has been practiced for many years, with communities remodeling themselves to create more roads, highways, and transportation networks, as well as new residential areas, industrial sites, and commercial areas. All of these factors work together to make society a better place.

Regenerated cities can occur on the earth's surface as well as through construction activity. Urban regeneration is currently taking place all over the world in order to build opportunities for people and provide a better quality of life for them and their families. With all of the benefits that urban regeneration brings, these countries will mature culturally, socially, and politically, allowing them to compete on a global scale.

6.2.2. Which Countries Are Well-Known for Urban Regeneration?

Some countries had urban renewal programs in place prior to the formation of European Union policies. This is the case of France, the United Kingdom, and even Spain, which has developed a new paradigm in the field of urban growth. Urban regeneration schemes have begun in Dublin, Ireland's capital, largely as a result of unequal building designs and inadequate building management.

6.2.2.1. Urban Regeneration in France

Cities in France have significantly higher densities than cities in Australia. For example, Paris has a population density of 10,000 people per square kilometer, which is more than five times that of Sydney, which has a population density of 1,900 people per square kilometer. Greater density and greater accessibility to public transportation are critical for effective urban regeneration. However, this is not the only reason for its popularity in France.

With the post-industrial world, new approaches to solving France's planning problems are developing. After the country's decentralization started in 1982, local governments have gained more power to enact development strategies.

At the same time, the number of urban stakeholders complicates decision-making. Legal obligations to meet with residents have risen since the 1990s. Regeneration programs must adhere to general planning standards while still allowing for some wiggle room to encourage the local community to participate.

For example, for Italian architects, city planning issues are and will be the primary focus in the near future. Cities in Italy, as in most other parts of the world, are in jeopardy due to a lack of resources and a deterioration in the condition of the put up-struggle building fund. According to the National Council of Architects, Planners, and Landscape Architects of Italy, it has become a true urban dehumanization; due to the lack of public spaces, land intake has reached an alarm stage, out of control growth of some vehicles, excessive intake and rate of strength, waste, and recyclable substances.

In general, the aim of city regeneration initiatives is to improve the city's impact by changing city lifestyles. These initiatives established strategic avenues and generated conditions for the strengthening and improvement of social capital, attracting monetary resources to transform public and private areas for the restoration of historic centers, housing districts, private residences, and the landscape. Suburbs must no longer be overlooked; their recovery must be the first step in addressing the cities and the environment's difficult conditions.

6.2.2.2. Urban Regeneration in Spain

Barcelona's urban regeneration has become a popular example of various cultural regeneration strategies within that adopted the approach of 'urban design, cultural planning, and innovative quarter approach' and included cultural thoughts in the renovation of areas alongside other initiatives within the environmental, social, and economic spheres.

Although we agree with other academics that the lifestyle and easy software of a completely unique 'Barcelona model' of city renewal is still debatable. Due to the 1980s period, extraordinary capabilities can still be seen in Barcelona's city growth. The city's creative combination of cultural activity and urban renewal has been bolstered by a distinct governance style built on strong citizen support, which few have dared to question.

The explanation for the lack of criticism is that the renewal of Barcelona has merged urban efficiency and social dignity with the goal of enhancing social cohesion and a "sense of belonging to the region."

6.2.2.3. Urban Regeneration in Turkey

Rapid population development, informal settlements, and buildings and facilities vulnerable to natural disasters are regarded as the most pressing urban issues in Turkey. Specifically, catastrophe threats cannot be ignored, as vast portions of various cities are at risk from earthquakes, flooding, and landslides, and have lost lives in the recent past. Urban regeneration is a critical planning method used by local and central governments to reduce disaster risk and create livable conditions for residents.

The Law on the Regeneration of Disaster-Affected Areas, also known as the Urban Regeneration Law, was passed in 2012. The law's implementation regulation outlines the basic steps of the urban regeneration process. The related agencies are empowered with different powers such as expropriation, confiscation, and changing the form and location of your land, making urban redevelopment projects critical in terms of property rights. As a result, urban regeneration tasks must be straightforward, understandable, and appropriate for all project participants. The laws and initiatives of various municipalities in Istanbul were examined in order to comprehend the urban regeneration technique.

6.2.2.4. Urban Regeneration in Iraq

The importance of urban renewal in improving the physical environment in Iraq is still underappreciated. Furthermore, the emergence of urban renewal projects in Iraq as a result of the decline of physical urban heritage in Iraqi cities raises concerns about the degree to which these initiatives have succeeded in identifying and seeking viable solutions to urban problems in terms of history conservation.

Recently, there have been attempts to establish a suitable appraisal strategy for local redevelopment initiatives, as well as to investigate potential solutions that can aid in reorienting city design strategies toward greater sustainability, mainly focused on evaluation of city layout aspects against a set of overall performance standards and indicators. By determining the relationship between urban design concepts and sustainable development goals in regeneration activities, urban design is central to the process of urban regeneration achievement.

6.2.3. Urban Regeneration and Climate-Friendly Development

Cities play a critical role in combating climate change and adapting to its effects because they contribute significantly to it and are severely threatened by its consequences. Since urban spatial policies have long-term consequences, they are critical for addressing climate change, and city councils may direct climate-friendly planning through such policies.

Spatial policies address concerns ranging from the regional to the individual building level, such as the promotion of compact cities, the development of green areas and water sources (retention and detention ponds, water canals, etc.), retrofitting existing buildings, infrastructure renewal, and increasing non-motorized and public transportation coverage. They can also be useful in achieving both climate change mitigation and adaptation targets at the same time. Green spaces, for example, reduce emissions by carbon sequestration and aid in the mitigation of effects such as heat stress, air pollution, and flooding.

The implementation of such spatial policies necessitates specific types of intervention in developed urban areas. "Urban regeneration" inherently includes such interventions, ranging from renewal to reconstruction, and can thus provide opportunities to implement climate-change-related spatial policies. However, most urban regeneration research has concentrated on community-based concerns, governance issues, and even sustainability, with little attention given to climate change.

Nonetheless, due to the ineffectiveness of the property-led approach in resolving problems of social justice and environmental sustainability, new priorities were incorporated into the urban regeneration model. The environmental benefits of developing existing urban areas were recognized in the 1990s, and redevelopment programs were seen as a way of addressing the three pillars of sustainability: economic revitalization, social justice, and environmental protection.

There have been recent efforts to link urban regeneration and climate policy, but these are still in their inception and will need further work to strengthen their theoretical and practical foundations. Current progress is limited to several examples of regeneration initiatives that are intended to contribute to climate change mitigation and adaptation, in addition to other goals.

The regeneration process involves various types of spatial interventions that can alter the shape and land use structure of cities, facilitating the introduction of spatial policies that resolve climate change.

The most successful method of action might be the productive use of inner-city lands. City governments will make the best use of brownfield and underutilized lands by urban renewal, allowing them to use a "grow-in" policy to concentrate the majority of new construction in existing urban areas, in the form of mixed-use developments.

By preventing urban sprawl and, in particular, reducing commuting time and distance, such a strategy can help achieve energy and resource efficiency. Furthermore, less energy is consumed for urban infrastructure activities in compact cities. We know that pumping water and storing wastewater consumes 30% of urban energy demand. As a result, the greater the area occupied by a population, the greater the amount of energy used in that city to provide water to and collect wastewater from buildings.

Buildings contribute significantly to greenhouse emissions due to energy demand for heating and cooling. Furthermore, low-quality buildings and those located in disaster-prone areas (e.g., floodplains) are the most vulnerable to climate impacts. Buildings that will be used in the coming decades are already being constructed in many countries. As a result, special attention should be paid to converting existing buildings into low-carbon, less fragile structures. As part of the renewal and renovation of inner cities, urban redevelopment could assist in overcoming such building-related challenges, either by retrofitting or renewing existing buildings.

6.3. SPORT TOURISM AND URBAN REGENERATION

The topic of regeneration has become a central element for urban policy and planning in much of the industrialized world over the last three decades. The definition of regeneration encompasses both physical dimensions, such as architecture and picture, and social dimensions, such as improving the quality of life for those who already live in target areas (Page and Hall, 2003). This spatial split was noted in the comments on urban regeneration delivery in Towards an Urban Renaissance: The Report of the Urban Task Force Chaired by Lord Rogers of Riverside for the British Government (DETR, 2000a).

There are several neighborhoods where restoration can only be accomplished by implementing comprehensive packages of interventions that address not only the physical landscape but also the economic and social needs of the residents. These areas include inner-city ex-industrial districts with significant amounts of derelict, abandoned, and underutilized land and buildings, as well as more densely populated areas, including many publicly-owned housing estates, that suffer from concentrated social deprivation.

However, if the social dimensions of regeneration programs are to be realized, it is clear that physical and social regeneration objectives must be combined in an integrated package. A study of urban renewal policies in the United Kingdom offers fascinating insights into the progress of such initiatives. The DETR-sponsored evaluation described 'regeneration' broadly as Area Based Initiatives (ABIs) primarily implemented by the Department of the Environment and/or DETR in England since 1990. These ABIs were implemented to solve cumulative, social, economic, and physical issues in underserved areas. Many regeneration initiatives in the 1980s and early 1990s were oriented toward land and property-led economic regeneration. Projects such as City Challenge and the single regeneration budget challenge fund (SRB), which put a greater focus on holistic regeneration by collaboration, as well as the more recent New Deal for Communities (NDC) projects, which aim significant resources at impoverished neighborhoods of 1000–4000 households, were among the initiatives. Although the evaluation found that the schemes improved some indicators, significant issues remain, especially in terms of assisting those who live in such areas. Indeed, it concluded that 'physical regeneration has, in many cases, played an important role in enhancing neighborhood identity and external appearance, as well as in bringing employment opportunities into the region-though the majority of jobs have not been secured by residents of deprived neighborhoods' (Figure 6.2) (DETR, 2000b).

Figure 6.2. Sport and tourism have long been regarded as urban revitalization tools.

Source: https://th.bing.com/th/id/OIP.aub0ahs2QuS8xK-3SKGKXwHaE6?pid=ImgDet&rs=1.

Sport, for example, is often seen as a means of overcoming social problems such as delinquency, and there is a rich therapeutic recreation tradition within leisure, recreation, and sport studies (Sugden and Yiannakis, 1982; Purdy and Richard, 1983; Hastad et al., 1984; Coalter, 1988; Glyptis, 1989; Robins, 1990; Tsuchiya, 1996; Witt and Crompton, 1996; Institute of Leisure and Amenity Management, 1999) The UK Policy Action Team 10 study, for example, proposed that sport (and the arts) would lead to "neighborhood renewal by improving communities' "efficiency" on four main indicators-health, crime, jobs, and education (Department of Culture, Media, and Sport 1999, p. 22). Sport, in particular, is regarded as playing a positive role in community growth: 'the strengthening of a community's social resources and processes through the development of contacts, relationships, networks, agreements, and events outside the household that residents themselves consider as making their locality a better place to live and work' (Thomas,

1992, p. 2). This conventional wisdom is frequently found as one of the cornerstones of government leisure and sport policies, especially in terms of regeneration strategies. According to the Wirral Partnership (2001), "sport supports the economy, the wellbeing of the community, and offers a constructive focus for individual and community motivation." Nonetheless, as Coalter and Allison (1996, p. 8) suggest in a review of the literature on sport and community development with respect to one of the UK sport measures of the 1980s (see Rigg, 1986; Deane, 1998): 'The lesson from Action Sport is that the transition from attempting to provide sporting opportunities at a local level for disadvantaged groups to the instrumental use of sport within communities is essential.' Indeed, the issue of the true social benefits that sports can bring to deprived areas is increasingly being called into question (Long and Sanderson, 1998), particularly given that unemployment and low income are at the root of social deprivation and urban vulnerability (Roche and Annesley, 1998; Coalter, Allison, and Taylor, 2000).

A number of studies have indicated that sports-related jobs will help with 'neighborhood regeneration' and community development (e.g., Department of Culture, Media, and Sport, 1999). However, there has been little research on the regenerative value of sport investment or the long-term benefits of sports-led investment initiatives to local communities (Gratton, 1999; Coalter et al., 2000). Present economic impact studies of local communities, for example (Henley Center for Forecasting, 1989), have simply measured the importance of the sports industry rather than addressing regeneration problems per se (Leisure Industries Research Center, 1997). Similarly, Coalter et al. (2000) report on Lincoln and Stone's analysis of the economic value of sport in the northern region, which concluded that "while many arguments are made for the contribution that this sector makes in terms of economic welfare, these are often based on assertion rather than concrete proof." There is a need for a more comprehensive assessment process to underpin sport support initiatives both nationally and in the field.' Unfortunately, such assessment processes are seldom accessible, and when they are, the results are often unsatisfactory in terms of the original goals for sports-generated economic growth (Hall, 1992, 2001; Crompton, 1995; Page and Hall, 2003). Despite these reservations, a variety of cities have embarked on large-scale sporting projects in conjunction with redevelopment plans, mostly in relation to the hosting of international sports competitions such as the Olympics (e.g., Sydney, Barcelona), the Commonwealth Games (e.g., Manchester, Melbourne), or even the World Student Games (e.g., Sheffield). The official City of Sheffield website, for instance, proudly proclaims itself as "Britain's first National City of Sport and boasts some of the best international sports venues in the country", including the Ponds Forge International Sports Center for swimming and diving and the Don Valley International Sports Stadium, both of which were built for the 1989 World Cup. In terms of its own regeneration plan, Manchester has recently positioned itself as a sports city.

The Eastlands Sport City site is a significant international development. Sport City, which includes the National Cycling Center, the North West Sports Institute, an indoor tennis center, and significant shopping and leisure opportunities, is built around the 48,000-seat Commonwealth Games stadium. Sport City served as the focal point for the 2002 Commonwealth Games, creating genuine new jobs and leisure opportunities for the local community. The extension of the Metrolink light rail system would strengthen connections to the city center and around the conurbation, growing work and social opportunities for residents (Manchester City Council, 2001).

Sports tourism is commonly regarded as a critical component for the success of sports-related regeneration initiatives, which are concerned with the reimagining of places in order to attract and sustain capital and people. Without a question, the "intangibles" of sport have a strong impact on decision-makers when it comes to the role of sport in economic growth (Baade, 1996; Whitson and Macintosh, 1996; Page and Hall, 2003). Baim (1994) analyzed financial data from 15 subsidized stadiums in the United States and discovered that, with few exceptions, the sole reasonable rationale for stadiums designed to accept or retain a sports franchise revolves around the external benefits (e.g., civic pride, tourism, and leisure time options) brought to an area. The costs to the taxpayer, on the other hand, varied greatly depending on the lease negotiations for each facility, with the greater the business size, the more likely that the market accrued external benefits. Nonetheless, because of the significant competition between medium-sized markets to draw sports teams, the per-capita subsidy appears to be higher, and the lease terms much more generous to teams from smaller cities (Whitford, 1993). For the 1995 season, the Los Angeles Rams relocated from Anaheim Stadium in Orange County, California, where they were losing $6 million a year, to St. Louis' new $276 million domed downtown stadium, which was subsidized by the city. In exchange, the Rams received a stadium lease with a rent of just $250,000 for the entire season, $13 million in moving expenses, half of game day expenses paid by the city, a separate practice facility, all revenue from boxes, club seats, and regular tickets, guaranteed sales of deluxe seats, the majority of advertising and concession profit, and

$27 million to pay off the lease in Anschutz Stadium (Sickman, 1995). Indeed, one recurring feature of analyzes of the economic impacts of sports stadiums and facilities, as well as the activities that fill them, is that their positive effects are often vastly overstated (Burns et al., 1986; Hall, 1992). Cromption (1995), for example, identified 11 sources of error contributing to overestimation in a study of 20 studies conducted in the United States. These errors included misrepresenting created employment (particularly in the service sector), focusing on total rather than marginal economic benefits, and ignoring opportunity costs. Similarly, Gratton (1999, p. 9) concluded that "while some data on the economic benefits of sports activities and sports tourism is available, many of the economic benefits to the local community have been poorly researched."

The potential attraction of sports tourism for urban areas is illustrated by Rawn's (1990) study of sports fans in Indianapolis. She refers to the transformation of Indianapolis from a manufacturing town to an international sports venue as 'smokestacks to stadiums: wealthy sports fans are a clean industry. Rawn cited a study of spectators at the 1989 GTE Tennis Championships in Indianapolis, where 36% had household incomes of more than $75,000 (compared to 4% of the Indianapolis population), one-third stayed in a hotel and went shopping, and 84% rate in restaurants. According to Rawn, such sporting events have contributed to the development of service jobs in Indianapolis, which have compensated for job losses in the manufacturing sector. However, Rosentraub et al. (1994) discovered that the sports strategy had little impact on development and economic growth in Indianapolis compared to other mid-sized cities in the region in a later analysis of the investments, policies, and strategy behind Indianapolis' US $172.6 sports-related economic development strategy (Rosentraub, 1996).

The Indianapolis strategy entailed significant municipal capital investment in 5 major facilities, the establishment of a National Institute for Fitness and Sports, and the hosting of seven sport governing bodies. However, as with many other types of urban development (Page and Hall, 2003), Rosentraub et al. (1994, p. 225) observed that 'it is very difficult, if not impossible, to completely disentangle the sports strategy from the no sports elements of the downtown development plan,' and that 'the overall sports strategy may have contributed to Indianapolis's image, the different or changed (1994, p. 233). Although the strategy 'generated a substantial number of service sector and hotel jobs' (Rosentraub et al., 1994, p. 237) and associated spin-offs from sporting event attendance, they calculate that sports-related jobs accounted for just 0.32% of all jobs in the Indianapolis economy (an increase of 0.03%) and sports-related payrolls accounted for less than 0.5% of total payrolls. Rosenbtraub et al. (1994, p. 238) came to the following conclusion:

- Without diminishing Indianapolis' prosperity and publicity, results of this magnitude are so minimal that it is reasonable to believe that, had the city concentrated on other factors, a greater economic effect would have been feasible. Given the small size of the sports industry and the low pay associated with the various service sector jobs generated by sports activities, sports are not a wise tool in which to organize a growth or regeneration effort.
- It is perhaps not surprising, then, that Coalter et al. (2000, p. 6.7) concluded in their review of event-specific evaluations that 'there is little evidence about the medium to long-term economic impacts of such sports event-led economic regeneration strategies. In general, there is a lack of adequate data on the regenerative impact of sports investments on local communities.' This is primarily due to the sector's significant economic benefits in terms of job creation, consumer spending value, investment opportunities, and the "footloose" existence of the industries within the sector. Nonetheless, despite this caution, sport and sport-related tourism remain important components of regeneration strategies. For example, the Northwest Development Agency (NWDA) in the United Kingdom states as part of its regional strategy, "NWDA will help to capture and sustain a long-term program of arts, sport, and cultural activities aimed at instilling pride and raising expectations." It attracts and assists in the retention of the most professional and talented members of the society-people whose leadership skills are needed to maintain the potential of every community' (2000, p. 34). More importantly for sports tourism, they later claim.
- Quality facilities for sports, the arts, and museums will help to develop the region's reputation and is part of the package required to attract and retain those with exceptional talent and abilities. These facilities are also popular with tourists and visitors. The NWDA, in collaboration with stakeholders, will recognize the strategic gaps needed to improve the region's role and reputation within the framework of a viable new investment action plan. The action plan would take into account the need for new facilities,

new construction, and upgrades to existing facilities (NWDA, 2000, p. 50).

- Without a doubt, facilities can generate jobs, especially during the construction process, and may generate some long-term employment, but in the case of event employment, most of it would be part-time, casual, and low-skilled. However, the degree to which such facilities have a strategy of training and employing people from the target area would be critical to their effective contribution to job recreation.
- The bid by Cape Town to host the 2004 Olympics is an example of the use of sport and sports tourism for urban regeneration in a developing world. Although the bid was unsuccessful, it was notable because it directly connected the hosting of a mega-event to development needs (Page and Hall, 2003). The Cape Town bid aimed to add a 4th pillar of "human growth" to the Olympic Movement's three existing pillars of sport, culture, and the environment. According to the Bid Book, "any aspect of hosting the Olympics can contribute to the upliftment and quality of life of the people of the city… we put particular emphasis on our deprived communities" (in Hiller, 2000, p. 441). Rather than focusing solely on urban renewal through the construction of new infrastructure and the promotion of the city, the Cape Town bid aimed to be transformative in both a social and economic context. As a result, the Cape Town bid brought two novel ideas into the position of Olympic host city. For starters, the Olympics will act as a catalyst for improving the social and economic conditions of traditionally disadvantaged communities. Second, they will work to redesign apartheid-era Johannesburg and forge new links between people and cultures.

The bid sought to achieve these goals by a variety of means (Hiller, 2000; Page and Hall, 2003):

- A transformational catalyst for change: using the Olympics as a tool to affect immediate short-term changes in the city's physical and social well-being, as well as longer-term effects.
- The development of facilities in deprived communities: seven of the 42 competition activity sites in the Cape Town region were designed for disadvantaged areas. However, 66 of the 77 proposed training sites were intended for de-

prived communities, thus providing a significant permanent resource in those urban areas most significantly affected by apartheid's legacy.

- Facilities as a 'kick-start' initiative: the construction of new facilities and the upgrade of some existing ones was viewed as a catalyst for neighborhood revitalization as part of a larger regeneration plan to draw new housing, retailing, and investment to deprived areas.
- Quality sports facilities supporting community sports programs: sport and leisure provision was seen as a way of enhancing the quality of life, as well as reducing crime and increasing community pride.
- A human resource opportunity: it was estimated that hosting the Olympics would result in the development of 90,000 permanent jobs in South Africa.
- Contribution to the stock of affordable housing: it was anticipated that the Olympic projects, including athlete and media housing, would make a small but substantial contribution to Cape Town's housing stock.
- Small business support: the bid expressly aimed to assist small businesses through an economic empowerment policy that provided 50% of its business transactions to companies from historically marginalized communities.
- Urban integration of the transportation system: 70% of transportation system construction funds were earmarked for initiatives that would specifically support marginalized communities by better connecting those areas to the larger urban structure.
- Community consultation-the Olympic bid group actively worked to include the community in the bid process through a number of processes such as local Olympic Steering Committees, a Community Olympic Forum, and a Strategic Environmental Assessment process.

• 'The concept of harnessing a mega-event to a wider urban agenda that goes beyond the concerns of finance capital, developers, inner-city reclamation, and the tourist city is a relatively new idea,' argues Hiller (2000, p. 455). This is particularly true given the preoccupation with winning IOC votes on a global scale while minimizing local costs and dissent.' However, it is worth noting that Cape Town did not win its bid (coming in third place in the final vote), and as a result, South Africa lost out to Germany in the bid to host the 2006 World Cup Soccer (although it did win the rights to the 2003 Cricket World Cup). As a result, the Cape Town Bid Company's argument that by awarding the bid to Cape Town, the International Olympic Committee (IOC) would have shown that the Olympic Movement was not "beholden to gigantism and commercial exploitation," but rather "devoted to the progress of all people and would therefore also provide opportunity to those already struggling for their place in the economic sun" (1996: Nonetheless, the Cape Town bid showed that the staging of sporting events can be used for the greater public benefit as well as the regeneration of cities as places of consumption, entertainment, and leisure (Hannigan, 1995; Page and Hall, 2003). 'When local people in the millions lack sufficient accommodation, food, and other subsistence needs, planning for a 'circus' when people need 'bread' would often seem inappropriate,' observed Hiller (2000, p. 455). However, the Cape Town bid introduces some new options for examination that could give mega-events new humanitarian urban value.' The Cape Town bid, like bids by Toronto to host the Summer Olympics (Hall, 2001), also indicates the increasing recognition that mega-event hosting, like perhaps all large-scale urban regeneration projects, must be seen as part of a broader social contract (Page and Hall, 2003). Indeed, one of the most important aspects of large-scale urban transformation initiatives, such as mega-sports activities, is that they often exclude participation in urban social democracy while requiring such large public investments that if they do not succeed as revitalization strategies, their real and opportunity costs are significantly altered (Page and Hall, 2003). Reflecting Law's (1993, p. 23) observation, 'urban policies are associated with both winning economic development for a city and rejuvenating the core areas, goals which may not always be consistent.'

- As a result, the future contribution of sports tourism to urban regeneration must be viewed in perspective. Sport and tourism, as well as their interactions, are widely seen as central to regional growth in municipal redevelopment strategies. Benefits are typically perceived in terms of infrastructure construction and the hosting of sporting activities that use such infrastructure, all of which led to jobs and the generation of a positive picture, which may then aid in attracting and maintaining resources, employment, and people.

- Sports facilities, as Coalter et al. (2000) concluded, can make an important contribution to the physical infrastructure of cities by having a social focus for a group and influencing people's perceptions of their neighborhood, as well as contributing to the quality of life of communities. Nonetheless, the contribution of sports infrastructure and sports tourism in the broader sense is subject to considerable discussion, hampered as it is by often problematic assessment and, as a result, often gross overestimation of the economic benefits of sport tourism in both the short and long term. In particular, the opportunity costs of investing in sport versus other redevelopment alternatives in terms of job creation are seldom considered. Sport, on the other hand, is incredibly difficult to argue against. Many people have an ingrained conviction that sport is good for them, that it makes them happier citizens, that it instills pride in their culture, and that it promotes a positive picture. This conviction, combined with a lack of criticism, means that many large-scale initiatives and activities will continue to be financed in terms of urban renewal, as they offer opportunities for politicians and public promoters to cut ribbons, unveil plaques, and be photographed with sporting winners. In fact, urban regeneration necessitates much more than sports and tourism to generate social and economic capital and create employment (Page and Hall, 2003). Investing in accessible and affordable education, wellness, and communications technologies, as well as a diverse job development strategy, is likely to have much greater long-term benefits for target areas than investing in elite, mostly commercial, sports clubs, facilities, and infrastructure. But maybe we will just ignore these topics and watch football instead.

6.4. SPORT-BASED REGENERATION STRATEGIES AS A MEANS OF ENHANCING THE IMAGE OF THE CITY TOURIST DESTINATION

Several cities in the United Kingdom, including Birmingham, Manchester, and Sheffield, have made concerted efforts and committed significant resources to bidding for and staging major sporting events, as well as building prestige sport facilities, over the last decade. Despite other proposed benefits such as urban renewal and local economic growth, an important reason for implementing these schemes is the alleged improvement in the city's reputation. Indeed, the use of sports initiatives as a means of improving city reputation is expressly acknowledged by the authorities concerned, as well as by a number of academic commentators. According to Loftman and Spirou (1996, p. 28), local officials prefer to concentrate on the city's reputation and future course rather than the comprehensive financial consequences of sports stadiums. As a result, recent sporting innovations in cities are often inspired and justified by a desire to create a new image for a community. Sport as a way of enhancing place-image, on the other hand, is not a new phenomenon. According to Reiss (1981), Los Angeles was a city that adopted a sport-based image enhancement policy in the early 20th century.

Following World War-I, a group of ambitious movers and shakers wanted to improve their city's image in order to encourage the expression of tourism, trade, and migration. The key to their strategy was the building of a massive outdoor sports facility that would host spectacular sporting events and festivals (Reiss, 1981, p. 50).

Despite such historical precedents, it could be argued that, since the mid-1980s, the idea of sport reimagining has been adopted with vigor and intent by a significant number of developed cities. Many of these cities, which have been heavily impacted by the restructuring of the international division of labor and the subsequent decline of their manufacturing sectors, have been forced to compete with one another as centers of consumption (Harvey, 1989). This has increased the importance of projecting a positive picture of the city to an external audience of potential tourists and customers. As a result, many cities have engaged in what has become known as location marketing (Madsen, 1992), which entails attempting to sell a place's image in order to make it more appealing to economic enterprises, visitors, and residents (Philo and Kearns, 1993, p. 3). Because of the media's current dominance and the prominence and attention devoted to sport, it has become normal for cities in the United States, and increasingly in the United Kingdom, to use sport as a medium for city image enhancement. While there are significant exceptions (e.g., the staging of the Commonwealth Games in Edinburgh in 1986), it appears that the policy of using sport events and prestige sport facilities to reorient cities has largely been adopted by 'industrial' cities. These cities are the product of industrial growth, and as

Law (1996) points out, their industrial character and appearance are a barrier to the development of their tourist industries (Law, 1996, p. 2). As Tim Hall points out, the term "industry" and the personalities and representations it evokes are extremely problematic for the promotion of cities in the form of the postindustrial urban economy (Hall, 1997, p. 216). According to Hall, it conjures up "a whole slew of negative, unfashionable images" (Hall, 1997, p. 216). In order to draw capital and citizens in this era of increased urban rivalry, these 'industrial' cities have attempted to create connections with more optimistic ideas (Harvey, 1989, p. 92). Sport's media attention, inherent prominence in modern society, and alleged positive connotations have resulted in the adoption of sporting programs by industrial cities as a means of image enhancement. As many observers have noted (Thorns, 1997; Harvey, 1989; Hall, 1997), the explicit goal of a significant proportion of city-imaging work is to improve how prospective urban visitors view the city. This section focuses on the tourism component of reimaging, with the goal of determining how the implementation of sport reimagining strategies affects the creation and structure of cities' images as urban tourist destinations. After relating the strategy of "sport reimagining" to the contemporary cultural context, we will investigate the characteristics of these initiatives that can enable them to influence city pictures. After that, an effort will be made to conceptualize city images in order to demonstrate how sport can be able to improve the city's reputation as a tourist destination. Finally, the consequences of this conceptualization will be addressed.

6.4.1. Sport, Tourism, and Image Enhancement

6.4.1.1. Sport Facilities

It is critical to recognize that, in addition to advertising campaigns and the staging of special events, the process of reimagining often includes the creation of new consumption spaces, which are often centered on glamorous 'flagship' or 'prestige' initiatives (Smyth, 1994; Loftman and Nevin, 1996). These flagship developments are large-scale, groundbreaking ventures that serve as a focal point and catalyst for visitors and media attention (Barke and Harrop, 1994). According to Harvey (1989), the development of these spectacular urban spaces provides cities with icons of urban dynamism, allowing the city to capitalize on conspicuous consumption in the midst of a sea of spreading recession. Sharon Zukin agrees and sees this method as a means of generating a vibrant symbolic economy from which a legible picture can be abstracted, thereby linking the city to consumption rather than output (Zukin, 1993, p. 45). Several cities in the United Kingdom have built sporting arenas as flagship or prestige programs. Examples include Birmingham's National Indoor Arena, Manchester's Nynex (now M.E.N) Arena and Velodrome, and Sheffield's Don Valley Stadium, Arena, and Pond's Forge swimming complex. These innovations, including the staging of special events, draw sport tourists to the area, but they are ultimately seen to have a more fundamental effect in terms of projecting a picture of transformation, allowing cities historically synonymous with industrial manufacturing to shake off 'the last traces of its 19th century self' (Westwood and Williams, 1997). According to Schimmel (1995), sporting arenas are viewed as icons of performance, urban machismo, and vibrancy, as well as multi-functional leisure facilities. This distinction between the functional use and perception of sport tourist items, as well as their use as icons, is critical in terms of image impact and will be discussed further in this section.

6.4.1.2. The Community Setting

Analyzing the basic essence of sporting interventions and their proposed effect on city photos above reveals that the use of such tactics can be seen as both an indicator of, and a platform for, such cultural changes witnessed by commentators in the post-industrial urban arena. In order to investigate the idea of consumer images of the city, it is particularly beneficial to link urban sport reimagining to literature and debates about changes in consumption patterns in contemporary society. David Harvey observes two significant changes in the realm of consumption. The first is the mobilization of fashion in mass markets as a way of accelerating demand across a broad range of lifestyle and recreational practices, including modern leisure and sporting habits (Harvey, 1989, p. 285). A second pattern is a move away from product consumption and toward service consumption. In practice, this means consuming entertainment, activities, happenings, and distractions. According to Harvey (1989), if there is a limit to the accumulation and turnover of physical products, cities should transition to the provision of ephemeral services in consumption. This is the mechanism by which the city transforms into a' spectacle.' The role of sport in this phase should not be underestimated, since sporting events are an important part of the spectacular city and, due to their popularity and visibility, can convey a positive picture of the city to a large and receptive audience of consumers. According to the argument, post-industrial cities have evolved into centers of consumption, play, and entertainment that are "saturated with signs and photographs to the degree that something can become portrayed, thematized, and made an object of interest" (Featherstone, 1991, p. 101). As Featherstone suggests in his use of the word "everything," the ways in which spectacular cities have used imagery to create symbolic or cultural capital have become increasingly diverse. Post-industrial cities have come to be known as centers of cultural activity, not only because of their links to the high arts, but also because of their ties to popular cultures. As a result, popular cultures, in which sport is a significant component, have come to be regarded as 'more real, a source of legitimacy, and higher up the symbolic hierarchy' (Featherstone, 1991, p. 106). Cities that have historically been synonymous with industrial manufacturing can therefore overcome perceptions of low cultural capital by using sport and other more common cultural forms to convey positive images to potential visitors. When contemplating city marketing campaigns or the intentional exploitation of a city's image, it is important to remember that much of the imaging is directed at the "better off" (Bramwell and Rawding, 1996) or "right sort of people" (Harvey, 1989). Without a thorough understanding of current changes in consumption patterns and the highly commodified sport 'industry,' it appears that sport is less fitting as an urban reimagining method. Sport in cities has traditionally been synonymous with the urban working classes, not the picture that would be considered to draw the "right kind of people" to cities that already have enduringly strong working-class reputations. However, with the growing gentrification of sport and the resulting misunderstanding of hierarchical customer preferences and desires, sport may be regarded as a more successful means of attracting the wealthy segments of the tourist market in the modern period. Whereas in previous cultural eras it would have been inherently contradictory to promote a city's sporting pedigree alongside more traditional forms of cultural capital, the argument here is that not only is this now an appropriate partnership, but one that can be effectively sold to a new market of more eclectic, rounded tourists whose tastes are not necessarily confined to the different ends of an ort. Along with the increased legitimacy of sport as a means of representation in an age where the line between high and popular culture has become increasingly blurred, there are other similar indicators that could point to the suitability of sport.

Sport as a 'tool' for imaging Sport reimagining is an appealing solution for communities aiming to attract the affluent customer but still trying to prevent the total abandonment and suppression of local culture and heritage. Sport has long been an integral part of life in the United Kingdom. This means that emphasizing this urban aspect as a tool for image enhancement could be a less contentious means of representation than other measures attempting to reorient the city towards being a place of leisure and play. For example, it may avoid the type of dispute that surrounded Glasgow's imaging scheme earlier this decade, which was centered on the city's designation as the European City of Culture (1990). Some people criticized the imaging campaign, claiming that the image displayed was 'not one that had sedimented through the years in Glaswegian consciousness, but one that promotes thinking about Glasgow in new terms, i.e., without relation back to external reality' (Boyle and Hughes, 1991, p. 221). While sport reimagining has been accused of being selective in its portrayal of a city, 'sanitizing the real working-class culture and cultural heritage' of cities such as Manchester and Sheffield (Boyle and Hughes, 1991, p. 225), there is little doubt that sport has played a crucial role in the growth and everyday life of these cities during the 20th century. This may imply that sport is embraced as a more' realistic' depiction of these cities, not only to appease sensitive people, but also, as will be discussed later, to portray a more' realistic' and therefore believable picture to a vital external audience. Whatever its position in terms of culture and intellectual capital, sport has evidently thrived in the post-industrial economy (Rowe, 1995). Sport has become an integral part of how cities have attempted to develop, espouse, and distribute images of the city as a revitalized center of spectacle, enjoyment, and play within this arena. As previously mentioned, the proliferation and dissemination of images is now an integral part of understanding the contemporary city. The city is right in the middle of this phenomenon. Cities are places where social representations are prominently displayed, where advertisement and promotion are most intense, and where the conspicuous act of consumption is most important (Lyon, 1994). Cities, on the other hand, have invested in their own images in an effort to turn themselves into consumption hubs. As a result, the connections between image and city are inextricably linked. Indeed, some argue that there is no such thing as a city, just an imagined world, and a picture that reflects a set of activities, relationships, and institutions. This is the city that we are interested in. The city of one's thoughts. Despite the impact and appropriateness of sport as a means of capitalizing on the contemporary value of urban imagery, it remains unclear what this definition of city image is, and more precisely, how sporting initiatives may be able to influence the way in which future customers (i.e., prospective tourists) create an image of the city in their minds. These questions are discussed in the subsections.

6.4.1.3. Deconstructing the Tourist Image

Image is regarded as critical in the development of tourism within a region. According to Fakeye and Crompton, tourist images are especially important because they modulate a representation of an area into the mind of a potential tourist and give him/her a foretaste of the destination (Fakeye and Crompton, 1991, p. 10). Within tourism research, it has been recognized that images in people's minds may have just as much to do with a region's success in tourist development as material and more tangible tourist products and resources (Hunt, 1975). As a result, in emerging urban destinations such as Birmingham, Manchester, and Sheffield, a primary consideration must be to project a positive destination image, in addition to developing attractive tourist products and infrastructure. Sporting programs can be valuable because they have the ability to achieve each of these goals. Despite academic commentators and municipal officials' protests about the critical role of sporting strategies in enhancing city image, it remains unclear whether and how sport reimagining can achieve this goal. Despite research on the impact of sporting initiatives on the external awareness of cities as visitor destinations (Ritchie and Smith, 1991), little or no research exists on the precise impact of those initiatives on tourist images. More precisely, the author is not aware of any studies that aim to describe these effects. The remainder of this section will thus outline how sport reimaging may be able to influence city image enhancement and make recommendations for effective measurement of these proposed impacts.

6.4.1.4. The Destination Image Concept

To assess the impact of sport-led strategies on city image, it is necessary to first define what is meant by the term "image" and how and why sport may influence this nebulous concept. The picture has traditionally been associated with the reconstruction of a scene or object in literature, painting, or film. This definition of the term is used in some academic fields (for example, the sub-disciplines of behavioral geography and environmental psychology), where it is common to refer to image as a mental reconstruction of a place, in this case the city, in a person's mind. However, as Raymond Williams (1976) points out, the traditional use of the term has given way to the concept of image as perceived reputation or character. Image has, in essence, become a jargon word in commercial advertising and public relations. It is argued here that it is perhaps most useful and widespread to use a combination of the various usages in order to explore the meaning of place image comprehensively, as the two meanings are not mutually exclusive. According to Echtner and Ritchie, image should be considered as perceptions of individual characteristics as well as more holistic experiences or mental pictures of a place (Echtner and Ritchie, 1993, p. 3).

Place images are created through a variety of sources, the most common of which are direct experience with the destination and important secondary sources such as the press, advertising content, television, radio, film, and literature (Gunn, 1998; Gartner, 1993). Gartner identifies autonomous image forming agents as one of the most significant types of secondary agents, which include independently generated reports, documentaries, movies, and news articles (Gartner, 1993, p. 201). According to Gartner, these agents may be the only image forming agents capable of dramatically improving an area's image in a limited period of time due to their perceived high credibility and market penetration (Gartner, 1993). This observation is important for sport reimaging since sporting initiatives produce and include autonomous image creation agents. Sporting connections are primarily transmitted through factual reporting and 'unbiased' news transmissions. While cities use deliberate sport-related advertising strategies, the majority of information obtained by potential tourists about sport in the city comes from sources that Gartner considers to be very reliable and unbiased. Significant sporting events and the construction of prestige sport facilities attract attention and press coverage, as well as media visibility from the coverage of the sporting events themselves. As a result, sport could become a credible, and thus successful, means of enhancing the image of the city as a tourist destination. According to Gartner, 'effective image change is dependent on an evaluation of currently held tourism images' (Gartner, 1993, p. 207). It would appear useful to know what images already exist of the destination when developing an effective image enhancement strategy, particularly because people can avoid conflicting details, or what Gartner refers to as 'cognitive dissonance' (Gartner, 1993, p. 205). Prospective tourists 'seek evidence that supports their views and try to disregard information that contradicts them' (Mayo and Jarvis, 1981, p. 35). This implies that, in order to be successful, imaging strategies must develop from the current images kept, rather than causing an immediate revolution of the destination image. This has far-reaching consequences for new urban tourist destinations. It may be unrealistic to believe that hyperbolic imagery of industrial cities declaring them to be enticing places to visit would be accepted by potential visitors. Cities in the United Kingdom, such as Birmingham, Manchester, and Sheffield, have a long history of hosting regular and one-time sporting activities and have become inextricably linked with high-profile teams and stadiums. As a result, associating prestige sport with these urban areas is not groundbreaking. Cities, on the other hand, may use 'major sporting events… to project a high-status image of the city through media coverage, which may help draw tourists' by reinforcing, renewing, and expanding these partnerships (Law, 1993, p. 94). It is also acknowledged that an individual's "current needs and expectations" play an important role in image creation (Ashworth and Voogd, 1990, p. 81). Pocock and Hudson (1978, p. 19) solidify the strong relationship between image and personal interests by defining image as "the sum of direct sensory information as interpreted through the observer's value predispositions," implying that a person's own interests, motivations, and values help determine the form of their place images. People's perceptions of cities are highly selective; what they want to view is directly linked to what they care about (Mayo and Jarvis, 1981, p. 28). If this is the case, the popularity of sports could mean that sporting initiatives can be more easily accepted into potential tourists' perceptions of the city. According to Whitson and MacIntosh (1993, p. 236), the popularity of professional sports and world events reaches far beyond a small number of people. According to the General Household Survey, 11% of adults had attended a spectator sport event in the previous four weeks in 1986, and this figure would rise to 13% in 1996, with a further 1% increase by 2002. (LIRC, 1998). In terms of the number of people interested in watching sporting events, the statistics for armchair sport spectating are even more promising. For example, nearly one-third of the world's population watched the Atlanta Olympics in 1996. (LIRC, 1998). Furthermore, the top 8 most-watched television shows in the United States are all sporting events (LIRC, 1998). The sheer magnitude of interest in sport, as well as its obvious popularity, can thus contribute to the penetration of sporting imagery. Mullin et al. (1993) also show a link between personal interest and image. According to the authors, the marketer's goal is to create a balanced or consonant relationship between the image of the city and the consumer's self-image so that there is overall consistency (Mullin et al., 1993). As a result, if a person is interested in sports, promoting a city through this medium may be especially effective. This is not only because a person would want to see the particular events/facilities in person, but also because the city's image and the person's self-image are congruent. Each picture enhances the previous one, resulting in a very positive relationship between prospective tourists and the area. A final point to make about the particular characteristics of sporting initiatives that may enable them to have a disproportionate impact on destination image creation is about 'imageability.' Kevin Lynch pioneered this idea, which refers to the quality of a physical object that gives it a high likelihood of evoking a strong image in any given observer (Lynch, 1960). In more recent discussions, it has been noted that authorities have promoted estheticizing or focusing on the visual use of public space (Zukin, 1998, p. 825, also Lash and Urry, 1990). In effect, the current trend is to capitalize on Lynch's notion of imageability by purposefully constructing urban features and spaces that have the potential to elicit a significant reaction from the observer. It is suggested here that sport stadia built as part of local economic development strategies have the potential to be substantially 'imageable' components of the urban environment and, as such, should be regarded as offering 'potent landscape features' (Stevens and Wootton, 1997). According to Raitz (1987, p. 5), sports stadiums provide cities with distinctive buildings that evoke a clear sense of location. Bale (1993, p. 3) agrees, stating that "it is the floodlights of the stadium, not the spire of the cathedral, that more often than not function as urban landmarks and points of reference." As previously mentioned, this visual aspect of destination image is critical and is often overlooked in tourism research (Echtner and Ritchie, 1991, 1993, MacKay and Fesenmaier, 1997). In conjunction with the previously stated factors concerning the penetrability, credibility, and success of sporting strategies, major event coverage transmits 'images of large scale, dramatic, and sometimes esthetically memorable stadia into our living rooms' (Stevens and Wootton, 1997). This means that the construction of sports facilities may be able to provide a city with significant visual icons that build a memorable and positive picture in the minds of tourists.

Conceptualizing the city image and city product Sporting initiatives may thus provide a reliable, practical, influential, and pervasive means of influencing how destination images are created. However, in order to better understand the ability of sporting initiatives to affect the structure of city destination photos, we must first conceptualize the city and examine how it is marketed and viewed by potential visitors. As a result, in addition to examining the impact of sport on the image creation process, it is also important to indicate how sporting initiatives can become a constituent of the type and nature of city image itself. According to Shaw and Williams (1994), one way to understand the various aspects of tourism in cities is to consider the urban environment as a commodity. Jansen-Verbeke (1986) attempted to conceptualize the city product, stating that cities contain primary elements (which incidentally include sporting activities and facilities), consisting of major tourist attractions, accompanied by secondary tourist elements including retail and catering facilities. The term tourism product, on the other hand, can be used at two distinct levels, one at the particular level and the other at the overall level-the entire experience (Smith, 1994). This point is supported by Lash and Urry, who observe that while tourists can consume a variety of tangible products, they often consume a location's core product (Lash and Urry, 1994, p. 215). As a result, the city as a holistic entity can be consumed when referring to the city as an urban tourism commodity, as well as when involving the consumption of specific features or functions of the city. As a result, it must be acknowledged that the city as a tourist product is purchased, sold, and imagined on various levels. This multi-level conceptualization of the city tourism commodity is critical when investigating the ability of sporting initiatives to improve the reputation of the city tourist destination. The city destination involves unique items such as the provision of sport in general, as well as individual attractions such as indoor sporting arenas on a smaller scale. These items can be purchased and sold separately from the holistic product. It is critical to realize that these various product levels both promote and produce distinct images that are critical to the urban tourism system. As a result, the value of picture in this multi-level city tourism product is paramount. Indeed, one might argue that the city tourism product exists only as a picture. This claim is especially prevalent in terms of the city's overall picture. According to Shields (1996), while we can talk of the reality of 'the community' as a thing or shape, the concept of the city as a holistic entity is essentially only a representation (Shields, 1996, p. 226). The term "city" is simply a label we use to describe a variety of activities, relationships, and spatial types. As a result, the material truth of the product is less important than how it is imagined. Hunt (1975) backs up this claim by stating, "Whether or not an image is a true reflection of what any given region has to offer the visitor, what is relevant is the image that remains in the mind of the vacationer" (Hunt, 1975, p. 1). The critical point is that, just as the city destination can be thought of as a multilayered phenomenon, possible tourist photographs of the city exist on various levels as well. The distinction between the city as a whole and specific product is equally important in terms of destination picture. As an example, consider a city that has a weak overall reputation but is nevertheless regarded as having outstanding sport facilities and bcing an appealing venue for high-quality sporting events. To fully understand this city's image as a tourist destination, it must be regarded on both levels. The city tourist product is thus essentially a collection of linked, but stratified product images, and it is argued here that in order to understand the effect of sporting initiatives, this product image stratification must be recognized. Echtner and Ritchie's (1991, 1993) findings about destination image are supported by the definition of the city destination appearing on different product image levels. Echtner and Ritchie contend that destinations are viewed by both individual characteristics and overall experiences. Using this framework, it is possible to propose that sporting initiatives can influence tourist perceptions of the city in two distinct but related ways. First, by developing the attribute-based image aspect-by developing perceptions of an important tourism product within the region. Second, sport reimagining has the potential to replace vague or negative views of the city as a whole with new holistic experiences and connections. The aim of the remainder of this section is to expand on this preliminary observation in order to create a context that can clarify how sporting initiatives are integrated into the picture of the city as a tourist destination. Echtner and Ritchie's (1991, 1993) conceptualization describes the various components present in photos of tourism products. What it does not do is explain how imaging strategies and other external factors can influence these components. Suggestions will be made below to further expand this critical consideration.

6.4.1.5. Dissecting City Images

As can be seen from the illustrated preliminary conceptualization, it is suggested that the representation of the city exists on two key levels. This simplified representation of the form of city images is influenced in part by the Gestalt psychological tradition, which states that the whole is greater than the total of the related parts of perception. As a result, while it is important to view city images as an amalgam of various facets of the city, the human mind often tends to consider places with a cohesive wholeness that distinguishes them from each other and the context against which they appear (Mayo and Jarvis, 1981). As a result, while the city can be viewed by its individual characteristics and roles, there is another degree of perception that considers the city as a whole. These two distinct, but inextricably linked, levels of city image are discussed further below, along with the implications of this fundamental conceptualization for the effect of sport reimagining strategies.

6.4.1.6. Enhancement of the City's Attribute-Based Image by Functional/Operational Image Enhancement

First, consider what is referred to as the attribute-based image aspect, in which the city is considered to have a variety of different functions and features. A tourist image of a city may imply that a city is envisioned as a location that provides certain recognizable tourist products. This image level is primarily a functional perception of the city in question, in which the individual imagines what characteristics and aspects might be present in a city and how the individual would use these attributes for their own particular purposes. This type of perception is referred to as operational processing of information by Appleyard, and it consists of the perception of 'goals, obstacles, and other elements relevant to purposeful action' (Appleyard, 1973, p. 109). Implementing sporting initiatives may have an effect on this city's image by improving awareness of a city's sport tourism product. As a result, the city's attribute-based reputation as a sporting destination may be improved, which may affect a specific segment of the tourist market. Urban sport visitors may be affected by imaging in a practical and organizational context as a result of the building of new sports facilities and the staging of sporting events.

Without devoting too much time to the issue, it should be stated that there are difficulties in defining the term "sport tourist." Hall (1992b) provides the most commonly used concept, which defines sport tourism as non-commercial travel to engage in or experience sporting events away from one's home range. However, care must be taken to avoid removing activities such as viewing sporting facilities or sporting exhibitions, which, despite their increasing prominence, do not appear to be recognized in Hall's description. For the purposes of this discussion, urban sport tourists are people who deliberately aspire to use, display, visit, or spectate at a city's sporting facilities during their stay. The sporting component of their visit may be the primary reason and inspiration for their visit, or it may be part of a larger package of tourism items in the city that they wish to sample. As a result, the ties between sport and the city are taken even more literally in this segment of the urban tourism market. Sporting programs are viewed as a distinct feature and draw of the region. The development of major sports facilities and the staging of high-profile events suggests that the city may grow an image as a sporting destination where tourists can go and experience urban sport tourism products firsthand. This clearly includes attending sporting events, but it can also include sport participation in both competitive and non-competitive contexts, as well as sport-related tourist attractions such as sporting museums and guided tours of specific stadiums. Several cities in the United Kingdom have attempted to improve their reputation in this way. According to Stevens and Wootton (1997), Sheffield has established a strong sport tourism product focused on its stadia infrastructure. This perception of sport as an 'experienceable' tourism commodity, and thus a means of promoting the city as a tourist destination, is evident in tourist literature created by the cities of Birmingham, Manchester, and Sheffield. This pattern is perhaps best exemplified by Marketing Manchester's promotional literature, which provides a hypothetical itinerary for the would-be sport tourist. As can be seen in this example, the city's new prestige sport facilities are being explicitly used to boost the city's reputation as a destination for experiencing prestige sport events. The expected picture impact thus relates to an improvement focused on the city's organizational or functional representation. This image enhancement can be distinguished from another method by which sport can improve the image of a city destination.

6.4.1.7. Symbolic Image Enhancement of the Holistic Image of the City

The holistic image aspect varies from the attribute-based image in that it includes overarching general impressions of the city as a whole. The attempt to simplify the urban environment into a single whole allows a tourist to organize knowledge and draw meaning from the dynamic and diverse contemporary community, which would otherwise be inaccessible to the imagination. As a result, the holistic picture is a critical aspect of the tourist decision-making process. According to Law (1996), some people travel to a city for a particular reason, however if we return to the Gestalt school of psychology, it could be argued that when considering the city tourism product, the amount is always greater than the pieces. According to Law, "when tourists are asked why they went to London or Paris, many do not say that it was to visit the Tower of London or the Louvre, but rather that they simply wanted to go there." (Law, 1996, p. 19) When considering the potential impact of sport reimagining on this holistic manifestation of city picture, it is argued that rather than merely improving the city's perception as a sporting destination, sport could have the potential to create a more optimistic holistic image of the city. When it comes to image enhancement for a city that has historically been synonymous with industrial manufacturing, the degree of city image can be affected by a general enhancement that replaces ambiguous or negative images with more positive holistic experiences (Figure 6.3).

'Inspired Sport'

Morning – Visit Old Trafford and the Manchester United Museum, together with a behind-the-scenes tour of the world's greatest football club.

Afternoon – Lunch at the Kilhey Court Hotel or De Vere Mottram Hall followed by a round of golf on their championship golf courses.

Evening – A sporting choice of superb venues, experience Manchester's Velodrome. Britain's national cycling centre and the fastest cycle track in the world or Europe's largest events arena, the Nynex arena in Manchester with its resident, world renowned ice-hockey and basketball teams.

Figure 6.3. Extract exemplifying the promotion of sport as an urban tourist attraction.

Source: Marketing Manchester Visitor and Convention Bureau (1998).

As a consequence, rather than any notion of sport as a part of the city, there could be an enhancement of the holistic picture based on abstract notions and symbolic representations. This form of enhancement is a connotative means of image creation because it includes what is implied or indicated in addition to sport tourism items, beyond their literal or explicit meaning. The implication is that sport tourism goods, like other consumer items, may have acquired the ability to accept a broad variety of imagistic and symbolic connections that supersede their original use value (Featherstone, 1991). As a result, the city has started to market itself in the same way that other consumer products do, not only as a practical commodity with specific characteristics and features that can be encountered through purchasing, but also through more abstract imagery associating it with positive ideas and emotions. Surfaces and styles have become more significant in the modern period, resulting in a 'designer' ideology. According to Jean Baudrillard, consumption should be understood as a phase in which only the signs attached to the goods are actually consumed, and thus resources (in this case the city) are valued for their significance rather than their use (Campbell, 1998, p. 103).

According to Strinati (1992), as people shop in supermarkets, they are just as interested in the packaging and design of the products on sale as they are in the goods themselves. If this is so, we might claim that we are gradually absorbing signs and pictures for their own sake. This may be especially relevant to sport reimagining, since, as Rowe (1995) contends, sport is inherently dependent on images and representations rather than the creation of use-valuable commodities per se. The implication is that using sport as an imaging tool may provide a city with an appealing picture, even if the recipients of the imagery do not perceive that aspect of the city operationally. Image enhancement is focused on a symbolic, abstract level of signs and symbols that affect how people visualize the city, regardless of their interest in spectator sport. Sport, rather than a particular function or feature of the city, provides the packaging, the picture. This point is especially applicable to sport reimaging because the proposed image enhancement does not have to be explicit in nature, as it does not have to explicitly apply to the city's functional dimension. The connotations associated with sporting activities and facilities are just as significant as the facilities and events themselves. Prospective tourist images of the city may be improved by the staging of a major event or the building of a sporting venue, even though they have no intention of sampling the city's sport tourism items. According to Hall (1997, p. 205), reimagining, and representing a city "relies on both material and symbolic capital." The construction of sports facilities and the staging of events has a material impact in terms of the provision of a particular tourism commodity, but the implications of the initiatives go beyond this immediate impact. The city may become a more appealing destination as a result of the positive perceptions resulting from the sporting initiatives, rather than the sporting initiatives themselves, which may boost how the city is perceived as a whole. These sporting connotations, according to Rowe (1995), 'emerge from the repeated affirmation of sporting ideals such as universalism, transcendence, heroism, competition, individual inspiration, and teamship' (Rowe, 1995, p. 138). Through this connotative method, it appears that reimagining the city through sport has the potential to influence the holistic aspect of the destination picture. The city of Manchester's sport reimagining could be a good example of this symbolic image enhancement. This city in the North West of England made two unsuccessful Olympic bids (for the 1996 and 2000 Games), but despite this obvious loss, used the bids as a focal point for reorienting the city's picture. While the mega-event never took place in the city and the majority of the planned sport facilities were never completed, the city's overall reputation was significantly enhanced as a result. In addition, simply being associated with sport in general, and especially the Olympic Games, changed ambiguous or negative impressions of the city into more positive perceptions. Rowe's evaluation of Manchester's Olympic bid agrees with this interpretation, believing that the city was eager to host the Olympics in 2000 because of "all its connotations of new era development" (Rowe, 1995, p. 137). Manchester has used the bid's legacy, such as images of sports facilities in promotional literature, to connote a new period for the region, highlighting the symbolic and abstract intent of much sport reimagining. The image enhancement in this case does not use the practical value of the bid in offering an enhanced sport tourism product, but rather on a more abstract, symbolic level, allowing the city's overall image to receive a much-needed boost. According to Rowe (1995), such examples "illustrate sports work as a metaphor, and its articulation with other traditions and principles constitute an appealing capacity to popularize often obscure conceptions of progress" (Rowe, 1995, p. 138). As a result, two major ways in which sporting initiatives can influence the image of cities have been established. The first is an organizational image enhancement focused on a better understanding of the city as a sport tourism destination, and the second is an influence on the overall image as a result of a connotative interpretation of symbolic and abstract imagery. Recent debates in critical sociology will advocate the importance of the symbolic over the practical, and the colloquial over the denotative when considering the current environment of consumption. However, the effectiveness of urban reimaging strategies may be dependent on capturing and using all types of image enhancement. The example of Manchester demonstrates how sport products can be used to improve both the functional and holistic aspects of a destination's image, using both connotative and denotative imaging. Despite the obvious risk of oversimplification, the various methods of image enhancement listed tend to provide a valuable way of conceptualizing the proposed effect of sporting initiatives on city photographs. However, in order to comprehend the significance of this conceptualization, it is important to discuss the consequences that arise from it. These will now be taken into account.

6.4.2. The Importance of Conceptualizing Destination Image

6.4.2.1. Research Implications

The recognition and understanding of holistic and operational image enhancement are critical because it is only through such conceptualization that the true effect of sport-led urban reimaging strategies can be measured. In order to fully understand the potential effect of a sport-based approach, research that seeks to assess image enhancement must view the tourism commodity at a variety of scales. City image analysis must recognize that while a sport-led approach can boost the image of the city as a whole, the policies introduced may also create perceptions of changes to a specific product (sport) within the city. This attribute-based image enhancement can occur even if overall perceptions are largely unchanged. Even if cities like Sheffield and Manchester are unable to create desirable holistic city images, they might be able to develop images and reputations as sporting destinations with an outstanding sport tourism product. The use of examples from Manchester to demonstrate both types of image enhancement provide proof that sport reimaging has the ability to improve both destination image components, thereby increasing the capacity to impact various sectors of the tourist market at the same time. The main point is that image analysis must examine these various strata of destination image in order to measure any proposed image enhancement resulting from sporting initiatives comprehensively. Furthermore, the complexity of any potential image enhancement in the creation of holistic images derived from symbolic imaging suggests that conventional image measurement methods must be made more versatile in order to allow for the understanding of such results. Although measuring attribute-based images with traditional structured questionnaires and rating techniques may be appropriate, measuring holistic image enhancement necessitates the promotion of less structured interviewing. Echtner and Ritchie state that "to fully capture the components of destination image-attribute, holistic, functional, psychological, common, and unique-a combination of less structured interviewing is required" (Echtner and Ritchie, 1993, p. 5). It is important that participants of any study be able to articulate their views on their own terms in order to clarify the meanings applied to specific programs and to investigate how sport reimagining is perceived. Despite the recognition and appreciation for such methodological implications in some contemporary geographical research, the relatively immature discipline of tourism has yet to adopt the methodological recommendations discussed here. While the discipline recognizes that symbols, signs, and holistic images are essential components of destination picture, the methodologies used to investigate the definition have not been changed accordingly.

Finally, in terms of methodological implications, an attempt has been made here to analyze how image enhancement strategies can impact the images kept by prospective urban tourists. More research is needed to link urban representation policies with the impact of these interventions on how the city is imagined. As a result, rather than focusing solely on how cities are perceived, more attention should be paid to how intentionally distorted imagery is viewed and interpreted by consumers. Despite the burgeoning literature on the promotion of city images and related improvements in urban shape, little to no note has been made of the ramifications of these changes to the external images of the city that exist in the minds of prospective urban tourists.

6.4.2.2. Marketing Implications

The above conceptualization can also aid future imaging of the city in targeting imaging strategies at particular segments of the tourism sector. The tourism literature widely agrees that place images must be planned to meet the needs of target audiences (Kotler et al., 1993, p. 99). In the case of sport reimagining, this inevitably necessitates attracting the interest of sport visitors. However, it is important to recognize that the most successful imaging could be one in which "the product is simultaneously marketed to different consumers with different needs and reasons for visiting" (Page, 1995, p. 216). Cunningham (of Proctor and Gamble) reaffirms this argument in his analysis of Newcastle-Upon-Tyne marketing. According to Cunningham, "you are searching for a multifaceted message that can appeal to diverse people with different agendas" (Armstrong, 1998). While sport facilities and sporting activities can draw a large number of people to the area, the amount of money spent on such initiatives means that, in order to be cost-effective, image enhancement should preferably influence a larger audience. The success of sport-reimaging can thus depend on the above-mentioned dual type of image enhancement, in which sport tourists and a broader audience can be affected. Hughes (1993) recognizes this claim in the case of Manchester. Hughes claims that the Olympic bid has aided the city's desire to grow leisure tourism by raising awareness and image of the city, which may directly stimulate sport tourism but may also "encourage a flow of visitors unconnected directly with the Olympics" (Hughes, 1993, p. 160). Recognizing the various levels on which city images exist allows for more successful marketing, guiding sport tourists to the sporting goods that the city has to offer, while using the symbolic capital provided by sporting initiatives to impact a wider audience-assisting in the 'circulation of images that impact climates of opinion and mentalities' (Zukin, 1996, 45). This colloquial imagery may result in a more positive image of the city in the minds of urban visitors, or at the very least pave the way for potential acceptance of more positive imagery by negating and ambiguous perceptions of the city. However, there is a risk that the beneficial effects will be negated unless the city's "multi-selling" is properly handled. It has been recognized, for example, that city marketing authorities prefer to take a composite image of the city and its place commodity when selling the destination to potential tourists (Page, 1993, 195). According to Bramwell (1993), cities' marketing campaigns are often overwhelmingly biased toward promoting a city's overall picture rather than targeting particular tourist classes. The risk for cities adopting sporting measures is that they can struggle to improve both the city's reputation as a sporting destination and the city's overall image to a significant extent. Sport reimaging campaigns have definitely been introduced in the cities of Birmingham, Manchester, and Sheffield as a way of marketing the city to an ill-defined general audience, with less attention being given to targeting unique imagery at the sport tourism market. This approach can imply that cities are missing out on the consumer segment that will be most affected by sport-based reimagining. Sporting initiatives may thus have the potential to improve the city's overall and as a sporting destination image. To capitalize on this, marketing strategies must be developed that revolve around the introduction of sporting initiatives aimed at particular market segments. This should include emphasizing the city's attribute-based reputation as a place to experience high-quality sporting action in spectacular arenas, as well as using sport to improve the destination's overall image. To reap the full benefits of this holistic image enhancement, cities will need to concentrate explicitly on the tourism market, rather than assuming that imaging techniques would automatically and concurrently project an appealing holistic image to developers, citizens, governments, and visitors alike.

Chapter 7

Sustainability and Sport Tourism Development

CONTENTS

7.1. INTRODUCTION

Tourism is one of the world's largest industries, with much of the expanding industry centered on pristine natural habitats including coastal and marine protected areas (MPAs). MPAs are gaining popularity among both international tourists and local residents. Tourism will support local communities and MPAs by generating revenue and creating jobs. Tourism, on the other hand, can endanger MPA resources by destroying habitat, disrupting biodiversity, and negatively impacting water quality, as well as endanger communities through overdevelopment, crowding, and disturbance of local culture. Furthermore, as tourist income "leaks" to outside operators, traditional tourism also does not help the local community. As a result, tourism has the potential to devastate the very resources on which it is based. Sustainable tourism, on the other hand, is purposefully designed to help local communities, value local culture, protect natural resources, guide a greater portion of income to the local community and MPA, and inform both visitors and local residents about the importance of conservation. Local municipalities, non-governmental organizations, and the tourism industry must all work together to create sustainable tourism enterprises that are both economically viable and beneficial to the local community. The chapter discusses the fundamental concepts of sustainable tourism, a new form of tourism supported by governments, environmental, and social institutions, and international organizations. It entails considering economic, environmental, and socio-cultural factors while planning and managing tourism. The chapter provides the historical context for the concept of sustainability as well as the major international events related to this subject. The chapter also discusses the detrimental consequences of tourism that can be avoided by implementing sustainable development principles.

Today, there is a lot of concern in the long-term viability of sports tourism. This concept is described as "the use of sports as a tool for tourism endeavors." This is a broad term. As a result, a sports tourist may be both an active participant and a passive spectator. There has been a significant rise in the number of sports tourism events over the last few decades. Indeed, sports tourists have a wide range of options for sports participation and spectatorship. As a result, many entrepreneurs see sports tourism as a significant business niche. Sports tourism is also critical for the economic development of tourism destinations and may have a positive impact on the host community. As a result, sports tourism is regarded as an economic activity with a significant impact on national, regional, or local growth. However, the growth of sports tourism may have a negative social and environmental effect on tourism destinations. The aim of this chapter is to address the interdependence of sports tourism and sustainability from various disciplinary perspectives.

7.2. CONCEPT OF SUSTAINABILITY

Sustainable development is often referred to as eco-development, self-sustaining development, or suspensory development. Economic development, environmental conservation, and social development are the three foundations of sustainable development. The word "social development" has recently been replaced by "socio-cultural development." This definition assumes a well-defined and deliberate relationship between the pillars, with the aim of ensuring intra- and intergenerational economic, environmental, and social balance (Figure 7.1) (Meyer and Milewski, 2009, p. 84).

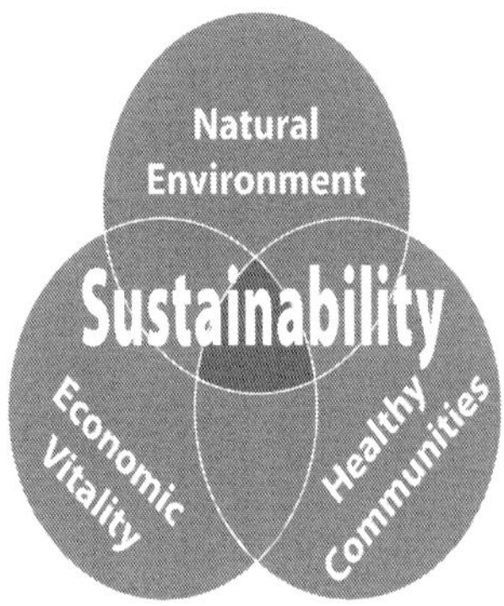

Figure 7.1. Concept of sustainability.

Source: https://th.bing.com/th/id/OIP.D2IALLXU4ouxcT3sVVOMLQHaDo?pid=ImgDet&rs=1.

The word "sustainability" is now used in the vast majority of regional development policies, and it also appears in the Polish Constitution. Sustainable tourism is characterized as all types of tourism operations, management, and development that maintain environmental, economic, and social integrity while also ensuring the preservation of natural and cultural resources. Sustainable tourism planning principles and practices apply to all types of tourism in all types of destinations, including mass tourism and niche tourism segments. Thus, implementing sustainable tourism concepts necessitates:

- Making the best use of environmental resources, which are critical to tourism development;
- Preserving basic ecological processes;
- Contributing to the conservation of natural heritage and biodiversity;
- Respecting host communities' socio-cultural authenticity, preserving their developed and living cultural heritage and traditional values, and contributing to inter-cultural understanding and tolerance;
- Ensuring long-term economic viability, offering equitable socio-economic benefits to all stakeholders, including secure jobs and income-earning opportunities, as well as social services to host communities, and contributing to poverty alleviation (UNEP, 2004).

To ensure broad participation and consensus building, sustainable tourism growth necessitates the informed participation of all relevant stakeholders, as well as strong political leadership. Attaining sustainable tourism is a continual process that necessitates constant monitoring of impacts and the implementation of appropriate preventive and/or corrective measures as required. Sustainable tourism should also sustain a high level of tourist satisfaction and provide visitors with a positive experience, raising their awareness of sustainability issues and promoting sustainable tourism practices among them.

Sustainable tourism addresses critical issues such as sustainable and socially responsible tourism industries, the ability for all people to engage in tourism, good job prospects in the industry, and the benefits of tourism activities for the local community. This necessitates the preservation of cultural identity as well as the incorporation of environmental conservation and cultural heritage services into tourism-related programs. Extensive efforts are being made at both the regional and tourism enterprise levels to put the ideals of sustainable tourism growth into action.

The International Ecotourism Society defines "eco-tourism" as "responsible travel to natural areas that protect the environment and sustain the health of the local people." It combines elements of both rural and cultural tourism. Ecotourism has its own set of values, in addition to adhering to the principles of sustainable tourism: It actively contributes to the protection of natural and cultural heritage, it engages local communities in planning, creation, and operation activities, and it benefits to their welfare, it provides full and interesting explanations for visitors about natural and cultural resources, and it is primarily intended for individual visitors as well as small organized groups. This form of tourism often involves outdoor activities such as hiking, mountain climbing, observing living beings in their natural environment, and so on. Furthermore, it may include cultural events. Ecotourism plays a significant role in education; it provides an opportunity to learn appreciation for nature and local culture, and for others, it provides an opportunity for self-reflection while being motivated by the beauty of the surroundings. Another distinguishing feature of ecotourism is the desire to support the local environment. This includes recruiting locals, sourcing local goods, involving locals in decision-making, and organizing tourist activities.

7.3. TYPES OF SUSTAINABLE TOURISM

There are several types of sustainable tourism, including: ecological tourism (ecotourism); green tourism; relaxed tourism; soft tourism; rural tourism; agro-tourism; community tourism; equitable tourism; solidarity and responsible tourism, among others.

7.3.1. Ecological Tourism

The term "ecological tourism" refers to the natural use (commercialization) of resources as tourism products in a sustainable manner. According to some authors, only tourism in natural reserves and national parks qualifies as ecological tourism. Hector Céballos-Lascurain, a Mexican, coined the term "ecotourism" in 1983 (subsequently, the director of the Ecotourism Commission of the International Union for Nature Preservation). It was a term developed to define a tourist travel focused on experience and exploration motivation, in a natural protected area, with a focus on education and tourist awareness of the environment.

7.3.2. Ecotourism

Ecotourism, also known as ecological tourism, is a form of tourism that focuses on preserving nature or interacting with endangered species. Ecotourism includes an important component of education and interpretation, as well as support for raising consciousness about the importance of natural and cultural resource protection. Ecotourism must have minimal environmental impact while also contributing to the well-being of local residents. Ecotourism became widely known in North America at the beginning of the 1990s as a reaction to the increasing interest in wild nature, and it represents a form of responsible travel in natural spaces that contributes to environmental protection and the improvement of local population life conditions. Ecotourism goods are designed with a special focus on nature, in areas with little human disturbance. In order to achieve sustainable growth, ecotourism requires respect for local sites and cultures. The intrinsic unity and human aspects of this tourism type were given special attention at the world ecotourism reunion in May 2002 in Québec (Canada). On the same occasion, the Declaration on Ecotourism was adopted, which expresses appreciation for the fact that this tourism type incorporates sustainable tourism principles in terms of economic, social, and environmental impact, and the Principles of Ecotourism were also adopted. The World Nature Alliance, on the other hand, claimed that the aim of ecotourism is to "admire, research, and appreciate the landscape, flora, wild animals, and any cultural characteristic of an area." For certain countries, ecotourism is more than just a sideline practice used to fund environmental protection efforts; it is a business that is the driver of the national economy and a source of significant revenue. Ecotourism, for example, is the primary source of foreign currency in countries such as Kenya, Ecuador, Nepal, Costa Rica, and Madagascar. There are still people who do not understand the importance of ecotourism and believe that it simply means constructing a hotel in the middle of nature, in a beautiful landscape, while ignoring the fact that this investment harms the local ecosystem. According to such people, ecotourism can first and foremost raise awareness of the beauty and fragility of nature, but they are unaware that they contribute to the deterioration of the natural environment by using air conditioning installations or pools, for example. Such activities are referred to as "green laundry" ("lavage vert") by tourism professionals, and it is believed that this form of activity conceals a traditional mass tourism that is marketed as "green" tourism. On a global scale, the United States is the largest source of ecotourists (over 5 million people per year), while the majority of other ecotourists come from Europe and the elites of a few Southern countries.

7.3.3. Green Tourism

Green tourism encompasses all types of tourism circulation that are in complete harmony with the natural and socio-cultural climate of the receiving region. It is also known as a collection of tourism forms promoted with the goal of improving tourism's social, cultural, and environmental effects.

7.3.4. Relaxed Tourism

Aside from the scope of nature preservation, which is a common trait for all tourism that aims at environmental compatibility as well as human health protection, this form of tourism has other goals: on the one hand, social goals (respect for local customs, traditions, social, and family structures), and on the other hand, economic goals (equitable revenue distribution, tourism offer diversification).

7.3.5. Soft Tourism

Soft Tourism distinguishes itself from the unnatural and impersonal modes of mass tourism.

7.3.6. Rural Tourism

Rural tourism is one of the most effective strategies for balancing tourism demands with environmental protection and sustainable development principles.

In a broad sense, rural tourism refers to vacations spent in rural areas, but this definition has proven to be somewhat imprecise, resulting in divergent opinions about the content and characteristics of rural tourism, ranging from the simple stay in rural areas to adhering to strict criteria related to tourist-consumer conduct, such as: consumption of agro-alimentary products from the hosts' homes.

In the current global economic context, rural tourism is described as the tourist valorization of:

- Rural spaces, natural resources, cultural patrimony, rural constructions, village traditions, earth products;
- Through brand products illustrative of regional identity, covering consumer needs for accommodation, nutrition, entertainment, and various services;
- To the end of local entrepreneurship.

If rural tourism encompasses all activities established while staying in the rural area, regardless of the type of accommodation unit, agrotourism is stricter, adhering to a set of holiday rules and taking into account the economic effects of tourism on households and rural establishments.

7.3.7. Agrotourism

Agrotourism entails staying in a household (boarding house, farm), consuming agricultural products from that household, and engaging, to some degree, in some basic agricultural activities. To meet the growing demand for participatory rural tourism types, the rural accommodation offerings have been enhanced by including theme holidays that invite visitors to explore nature (walks or horseback rides), local or regional gastronomy lessons, wine, and traditional food product sampling, and so on.

7.3.8. Community Tourism

Community tourism focuses on engaging local communities in tourism growth that is localized and built for their benefit: they design and operate the accommodation structures as well as the local services provided to tourists. Local populations have full leverage over tourism-generated revenue, with a large portion of the revenue going toward improving the community's living standards, with particular emphasis on environmental stewardship and local population traditions. This type of tourism growth is often coupled with the development of manufacturing activities, such as agricultural product transformation or handcraft workshops, the products of which are mainly sold to tourists.

7.3.9. Equitable Tourism

It represents a concept that involves applying within the sector the principles of equitable commerce. While less established than equitable commerce, equitable tourism is being implemented by an increasing number of tourism organizations and enterprises, with the goal of ensuring that communities in tourism destinations receive an equitable share of tourism income, as well as tourism reconciliation with community sustainable growth. For example, equitable tourism entails:

- Tourist holidays realized through consultation with local associations, consortia, and local population, collaboration, and joint effort for holiday elaboration;
- A fair remuneration of local partners, through total transparency of the tourism product price mechanism;
- Commitment to a long-term relationship with local populations, with a single goal: the improvement of the environment.

Tourists must also be educated about the implications of their vacations in order for them to be responsible. In general, equitable tourism refers to a set of parameters that focuses on resident and environmental respect, interactions between tourists and locals, and the long-term viability of tourism progress for local communities. Tourism projects are therefore developed by societies, or at least in collaboration with them. The communities effectively engage in the evolution of activities established by visitors inside tourism destinations, with the option to alter, re-orient, or even discontinue them.

7.3.10. Solidarity and Responsible Tourism

According to the International Forum for Solidarity Tourism, held in Marseille, France, in 2003, solidarity, and responsible tourism is a social movement that seeks to regulate and valorize the tourism economy for the benefit of local communities at destinations, in a territory development intercession. This intercession is constructed entirely from the society's human, cultural, economic, and environmental capital, which shape the life structure of local communities. This tourism form involves the responsibility of all the actors involved: the host community, intermediaries, and visitors, with the responsibility centered on respect for local customs and values, as well as a certain fair allocation of the produced revenues. As a result, equitable tourism advocates proper remuneration for various entities involved in the development and commercialization of tourism products, especially for the local population. One could expect that visitors will agree to pay more for services, recognizing that the difference will allow basic service providers to be paid more. Sustainable tourism services, like consumer products, are also dependent on labels that certify their quality. Solidarity tourism creates a dialog, solidarity, and mutual aid relationship between tourists (mostly from developed countries) on one hand and their hosts from developing countries on the other. With solidarity tourism, nongovernmental organizations (NGOs) from developed countries form alliances with NGOs from developing countries, which then work with local communities.

This form of tourism is still relatively unknown to the general public and suffers from a lack of regulation, efficiency, and collaboration. Efforts to promote equitable and social tourism are focused on two strategic elements:

- As compared to larger players that seek to maximize income, the communities of the South are the weakest link in the tourism chain, as competition significantly reduces their benefits; and

- A growing number of tourists from the North, as well as many NGOs and other citizen associations, are ready to support this process of local society consolidation and to maximize economic results to the direct benefit of these populations, promoting discovery while respecting identity and values.

Responsible tourism demonstrates that, although dependent on consumer and commercial principles, there is an alternative to mass tourism product consumption. Responsible tourism, as an experimental field for future tourism, valorizes the specificity of places, patrimony, and local culture, and tests cross-border good practices, such as benefit distribution among local populations at destination. Responsible tourism encompasses the following forms of alternative or advanced tourism: ecotourism, solidarity tourism, community tourism, and tourism "in favor of the poor." Integrated tourism is described as local tourism developed in rural areas, inbound tourism desired and developed by residents, and tourism for meeting and interrelationship growth. The character of confusion corresponds to a relationship that is established among various local players in order to provide the service. The contribution of these tourism forms to local population development is a common feature. Others apply to policy, operational modes, performance, business niches, and human and financial resources devoted to growth activities. Traditional tourism (classic) has received a lot of flak. For example, developing countries cannot "master" tourism flows, which are generally controlled by international groups from industrialized countries, and developing countries' external balance of payments is frequently negative, as foreign currency income is outpaced by import product demand. From a social standpoint, tourism jobs are typically low-paying, seasonal, and offer few opportunities to obtain real qualifications. Tourism, it is said, frails the social tissue, shakes cultural foundations, and deepens social inequalities by implementing unsustainable consumption patterns. Gas emissions from transportation have greenhouse effects, leading to environmental disequilibrium, and their impacts on Earth include deforestation, overexploitation, and landscape destruction. From the standpoint of local and foreign constraints, as well as its reliance on public sentiment, fashion trends, market patterns, and so on, the current tourism offer has a number of flaws. As a result of the high growth in frequency, one of the dangers for responsible tourism is the saturation of locations, areas, neighborhoods, and people in charge of tourist reception. Another risk is the tendency to create niches, or extremely narrow client segments, as well as to practice destination, product, territory, or client elitism.

Some tourism associations and tour operators have publicly committed to changing their practices and supporting local projects from destinations in order to assist residents in developing. When it comes from the associations business, this attitude is considered natural, but it is more shocking when it comes from tourism market operators. To get out of the uncertainty of the offer (solider or responsible), it is important to reach out to the client, since this offer corresponds to a specific demand, although one that is still limited in volume, in both Europe and North America, and brings benefits to those who coordinate these types of actions. In the same way that rural tourism started as a source of local growth and evolved into an offer for an increasingly diverse clientele, solidarity, and responsible tourism would be able to "enrich" its promoters and organizers.

7.4. NEED FOR SUSTAINABLE TOURISM DEVELOPMENT

Until the beginning of the last decade, tourism was regarded as a lucrative industry with no apparent constraints to development, few barriers to entry into the market, almost universal government support, and, for the most part, few effective regulatory requirements to consider the environment.

Commercial corporations, big and small, follow Dawkin's self-interest theory. They do not make drastic improvements to their corporate practices, which may be due to exhortations or good intentions, except in reaction to external conditions that cannot be prevented or to gain a competitive advantage.

International tourism has ushered in a period in which the prospect of making fast money by exploiting what were previously thought to be freely accessible natural resources has dazzled the eyes of government and business leaders, as well as many local residents.

Since tourism growth faces numerous economic, social, ecological, and political constraints, long-term solutions are needed to address these issues.

7.5. PRINCIPLES OF SUSTAINABLE TOURISM

The principles of sustainable development are rarely implemented in tourism enterprises. That is because the economic measure of tourism companies does not include (or only includes to a small extent) the social costs and costs associated with environmental protection. In practice, the most popular strategies are those that are low-cost and help to minimize costs (e.g., energy savings in hotels, recycling), allow for better positioning of brand companies to differentiate themselves from the competition, and elicit a positive reaction from consumers.

1. ***Using Energy in a Sustainable Manner:*** The protection and fair use of resources-natural, social, and cultural-is critical and makes good business sense in the long run.
2. ***Reducing Waste and Overconsumption:*** Reducing overconsumption and pollution saves money on the costs of repairing long-term environmental harm and improves tourism efficiency.
3. ***Preserving Biodiversity:*** Maintaining and fostering natural, social, and cultural diversity is critical for long-term sustainable tourism and builds a strong foundation for the industry.
4. ***Incorporating Tourism into the Planning Process:*** Tourism growth that is incorporated into a national and local strategic planning system and conducts environmental impact assessments improves tourism's long-term viability.
5. ***Assistance to Local Economies:*** Tourism that promotes a diverse range of local economic activities while accounting for environmental costs and values preserves these economies while also avoiding environmental harm.
6. ***Involvement of Local Communities:*** 'Local communities' complete participation in the tourism sector not only benefits them and the community in general, but it also increases the quality of the tourism experience.
7. ***Consultation with Stakeholders and the General Public:*** If the tourism industry and local governments, organizations, and agencies are to work together and address possible conflicts of interest, they must consult.
8. ***Staff Training:*** Staff preparation that incorporates sustainable tourism into work practices, as well as personnel recruitment at all levels, increase the efficiency of the tourism product.
9. ***Responsible Tourism Marketing:*** Tourism marketing that offers complete and responsible information to visitors' increases appreciation for the natural, social, and cultural ecosystems of destination areas and improves consumer satisfaction.

10. ***Conducting Research:*** Continuous market research and monitoring using efficient data collection and analysis are needed to help solve issues and deliver benefits to destinations, the industry, and customers.

7.6. SUSTAINABLE TOURISM DEVELOPMENT

Tourism is an activity that can help a country achieve a high level of economic and social growth. However, in order to achieve this goal, tourism growth must be focused on proper planning and well-managed so that the beneficial effects do not turn into unfavorable consequences for people and the community. Anarchic and speculative tourism growth may have serious consequences for conservation efforts and the wise use of natural and cultural resources. In these conditions, sustainable tourism can be regarded as one of the big stakes of tourism policy at the moment. In reality, in addition to beneficial economic aspects, tourism growth often has unfavorable implications, most notably the possibility of a significant deterioration of the natural environment, lifestyle, and cultural and social patrimony at the destination. This risk is linked to the phenomenon of overexploitation as well as the high density of tourism flows to specific destinations (Figure 7.2).

Figure 7.2. Sustainable tourism development.

Source: https://th.bing.com/th/id/OIP._BuwOHktRU7ut0kKD7nFGQHaDk?pid=ImgDet&rs=1.

The latest tourism growth strategy in an increasing number of countries aims to orient this sector toward a tourism that combines environmental stewardship and the creation of new jobs, as well as holidays that are affordable to everyone (by most people). Tourism planning must be based on sustainability principles, be economically sustainable in the long run, and be ethically and socially equitable for the local population of tourism destinations. Tourism must contribute to sustainable development by integrating into the environmental, cultural, and human environments. It must also value the delicate balance that is characteristic of many tourism destinations. Geographic concentration of foreign tourist flows threatens tourism growth to the point that the tourism sector's foundation may be jeopardized. The case of those areas with a high tourist concentration, where restructuring problems must be resolved, is well known: unsuitable ecosystems, circulation issues, devastation of the environment (particularly beaches and seashores), phonic pollution, and general natural environment pollution Even if they only receive a limited number of international visitors, developing countries are not immune to unfavorable effects and the possibility of tourist rejection, even before tourism becomes one of these countries' economic development drivers. Because of these factors, sustainable tourism management is a critical component of success in incorporating tourism into development goals. Inadequate tourism development has the potential to seriously harm the climate and social equilibrium. In such cases, tourism may result in the deterioration of natural and cultural patrimony, as well as the alteration of local population traditions and lifestyles, to the point where tourism is not tolerated in some areas due to rivalry between locals and tourists for access to infrastructure and public equipment. A focused sustainable tourism would be critical to the sustainability of tourism sustainable growth. In reality, tourism has a number of economic characteristics that set it apart from other industries. For the majority of developing countries, available resources for tourism growth are often inadequate and cannot be mobilized all at the same time. This includes choices, particularly for tourism enhancements and general and specific infrastructure, such as road access, potable water, electricity access, and telecommunications. The choices cannot be chosen except as part of a well-planned tourism growth strategy. The key issue in this regard is aligning tourism management with long-term growth goals. Traditional planning does not seem to be able to respond to the objectives of tourism growth any longer, as this often involves selecting one or more areas that, if effective, will become highly concentrated tourist areas, which is counter to the objectives of sustainable tourism development. Tourism growth management should be able to play a new role in these circumstances, that of sustainable tourism development without achieving an excessive concentration of tourism flows and the negative consequences of this concentration, a situation that may inevitably lead to a rejection of tourism by the local population. This is the scope of the current challenge-the revision of the priorities and, in particular, the practices of tourism resource management that should be capable of allowing the solution of specific problems of financing the tourism sector, while keeping in mind the growing competition among destinations on the international tourism market. To summarize, there are various types of sustainable tourism, some of which are unknown to the majority of visitors, but sadly, also too many tourism organizers or service providers. Consumers of experienced tourism goods have become more attentive to environmental concerns and the sustainable nature of tourism activity. They anticipate that new deals will be made in response to these requests. It is the responsibility of tour operators, first and foremost, to adequately respond to these requests through new offerings.

7.7. POLICIES TO PROMOTE COMPETITIVE AND SUSTAINABLE TOURISM

The tourism industry faces the major and multi-faceted challenges that demands an integrated approach to policy development through many government departments. The role of Coherence and consistency are vital in the application and design of policies between all phases of government to ensure that tourism policies are in effect.

7.7.1. Key Challenges Facing Tourism

Under current circumstances, the tourism industry faces a range of critical challenges that will affect its long-term growth and progress, as well as set the agenda for public policy action.

These include:

- Globalization and changing markets;
- Economy-wide impact of tourism;
- Climate change and sustainability;
- The knowledge economy;
- Human resources; and
- productivity and competitiveness.

7.7.2. Globalization and Changing Markets

Globalization refers to both the process by which economic markets, technologies, and communications become more international over time, as well as a wide range of social and other issues. For example, the increased liberalization of international trade regimes and the elimination of barriers to all forms of international exchange enabled the current global crisis to spread rapidly among the world's financial institutions, causing the worst global recession in 60 years. These same liberties also facilitate the rapid spread of global health threats such as SARS and H1N1 flu, promote consumer empowerment through access to global knowledge via the Internet, and ensure that developments in global commodity prices due to rapidly increasing demand from emerging markets now occur almost instantly. Tourism must adapt to the new global economy. It must deal with the threats to competitive advantage posed by the rapid dissemination of information and the economic impact of globalization. Tourism, of course, is both a major player in the globalization process (via the rapid expansion of new destinations, demand, and markets) and is heavily affected by it. For tourists, it has facilitated cross-border travel by lowering trade and travel barriers as a result of transportation liberalization. Rapid economic growth throughout many developing countries, liberal economic policies, dramatic changes in transportation efficiency and cost, and a growing global middle class (now estimated to number more than 2 billion people) with higher living standards have introduced a new dynamic in international tourism flows.

Underpinning tourism growth has been an increase in international trade and investment that has outpaced global GDP growth for the majority of the last decade. Economic growth in the OECD region has benefited dramatically in recent years from the dynamism of large non-member economies, particularly in Asia. Non-member economies are expected to account for 60% of global economic production (in Purchasing Power Parities terms), half of global trade (at current market rates), and one-third of nominal cross-border asset and liability assets by 2025 if the current rate of globalization continues. Globalization is the current economic and social reality that tourism must work within. The tourism industry, by definition, is resistant to the changing market realities brought about by globalization, but it can also be a major beneficiary of globalization if the appropriate adaptation strategies are implemented.

Tourism markets have been evolving and will continue to shift in tandem with these global trends. The exponential growth of outbound tourism from new markets, especially China, Russia, and India, is changing patterns of travel flows and demand. These necessitate the acquisition of new marketing and customer service skills, as well as the creation of suitable products. Sequentially, changing social values, lifestyles, and developments are causing profound social changes in developing countries, which are progressively reflected in changes in tourism demand. These changes can be seen from the increasing fragmentation of tourism markets and the development of new niche markets (e.g., retired travelers in developed countries; health tourism in developing countries). Holidays for singles, retired people, and money-rich/time-poor short-break travelers, for example, are emerging as major market segments in OECD economies, in comparison to the original annual family holiday. Patterns for adventure, active, and higher involvement perspectives, extravagant, and luxury travel, a search for unique experiences, and a demand for authenticity are among the other emerging niches. Customers are becoming more adventurous, eager to try new products, foods, and attractions, and are unwilling to give a second opportunity to destinations that provide subpar products or services. With increasing business mobility and the growth of global enterprises with growing expatriate workforces, globalization is likely to continue to stimulate the development of business travel. Meetings, discounts, conference, and exhibition travel will continue to expand rapidly. Business travelers who are short of time will aim to merge business and leisure trips. Such changes must be considered when developing tourism policies.

7.7.3. Economy-Wide Impact of Tourism

Tourism stimulates activity in almost every industry across the economy. The process by which the consumer (visitor) arrives at the product (the destination), expends identifiable tourism products, and accrues normal day-to-day living costs in that destination is distinct among traded goods and service markets. The diverse nature of the products as well as services consumed by tourists raises issues that are not found in other product markets. Furthermore, tourism has a social impact in areas as diverse as crime, health, traffic congestion, land, and other prices, and urban amenities. Tourism, according to research, provides critical forward and backward linkages with other sectors of the economy (see Tourism in OECD Countries 2008, Section 2B). This demonstrates the extent to which tourism has a diverse set of stakeholders and demonstrates the need for governments to address both horizontal (across agencies responsible for transportation, infrastructure, regional development, immigration, and customs, education, and training, and so on) and vertical (from the national to the provincial, regional, and local levels of government) issues. Many industries are defined as "tourism characteristic" or "tourism connected" in tourism satellite accounts (TSA).

Visitors' spending goes directly to a diverse range of businesses located in the tourism destination, many of which do not identify with the tourism industry at all and may have little understanding of the extent to which their sales are generated by tourism demand. Furthermore, the industry's composition is overwhelmingly skewed against SMEs operating in highly competitive markets.

Such businesses normally lack the capacity for long-term planning, analysis, and strategic growth. In most cases, they manufacture and sell only one portion of the tourism product. The tourism industry's complex web of stakeholders creates a fragmented structure that governments may find difficult to serve through general policy measures. As a result, governments are increasingly recognizing that tourism necessitates a complex set of mutually supporting infrastructure, policy, and planning decisions in order to realize the broad-ranging nature of tourism's benefits and manage potential costs. Although tourism remains primarily a private sector operation, some governments recognize that they have a significant role to play in resolving market failures caused by tourism's fragmented SME-based structure. It also becomes clear that tourism can provide broader economic and social benefits to destinations, especially by assisting in the sustainability of local and regional communities. To maximize economic and social benefit from the tourism sector, a "whole-of-government" framework for tourism policymaking should be encouraged. The impact of globalization on tourism demand, transportation, information, and vulnerability to external shocks has highlighted this. As a result, there is now a new imperative for the promotion of dialog, cooperation, and collaborations among a wide range of stakeholders in government, the private sector, and local communities.

7.7.4. Climate Change and Sustainability

Climate change and global warming pose a monumental threat to the global economy and culture, with far-reaching consequences for travel and tourism, whether or not policymakers take successful mitigation or adaptation steps. Tourism is expected to account for up to 5.3% of global anthropogenic greenhouse gas emissions, with transportation responsible for approximately 75% of this. On the plus side, growing awareness of the relevance and opportunities of green jobs-the promotion of lower-carbon activities and more efficient energy use through government policies-provides tourism with an opportunity to adapt its operating practices and participate in the expansion of "green-collar" jobs, which can be seen as a major growth opportunity for global employment.

Tourism is especially vulnerable to the expected direct effects of climate change, which include sea-level rise, changes in ocean currents, increased rates of glacial melt and loss of snow cover and permafrost, higher temperatures, changes in precipitation patterns, and an increase in the frequency or severity of extreme weather events. Such effects can result in biodiversity loss, negative impacts on the natural and built environments, and negative impacts on tourism-related infrastructure.

Other consequences may include heat stress's health and safety implications, the spread of tropical diseases, and the possible loss of main brand icons and photographs. Coral reefs, ski resorts, and island beach and dive resorts are particularly vulnerable. The direct effects of some impacts of climate change can be expected to produce winners and losers in the tourism industry. Global warming could make some destinations more appealing to tourists by making previously inhospitable climates more appealing. The widespread existence of the expected impacts, on the other hand, indicates that many other destinations may face severe and costly consequences. Changes in consumer behavior in response to climate change, the impact of increased voluntary or mandatory carbon offset schemes, potential modal shifts in transportation use, and demand for environmentally responsible behavior all have the potential to pose threats or challenges to developed tourism destinations and enterprises, with particular impacts for longer haul travel. Changes in the relative travel and tourism competitiveness as a result of new mitigation or adaptation measures implemented in any one destination or economy could result in rapid shifts in the direction and intensity of international tourism flows.

Climate change and its consequences for tourism necessitate that the tourism industry become even more concerned with the overall environmental sustainability of its own actions. The world that visitors want to visit has a major impact on the feasibility and attractiveness of tourism destinations. Climate change has the potential to dramatically widen and escalate environmental threats. This comes at a time when public interest is high and there is a need for stronger environmental safeguards. In the face of these mounting stresses, tourism remains vital to the economic well-being and, in many cases, the survival of many local and regional communities, because tourism occurs at the destination level, which is often associated with a local community. The relevance of tourism to the economic wellbeing of local people is evident in such locations.

The willingness of a wide group of stakeholders, across levels of government, industry, and local communities, to collaborate to adopt appropriate measuring and regulatory instruments for maintaining community- and environment-friendly results is critical to a destination's sustainability. Governments are increasingly attempting to incorporate social, cultural, and economic goals into the preparation and implementation of effective tourism and environmental management strategies that address longer-term perspectives. This emphasizes the vital value of a "whole-of-government" approach to tourism policy, which recognizes not only wealth and economic growth goals, but also environmental, social, and cultural concerns.

7.7.5. The Knowledge-Economy

Globalization, particularly its impacts on information flows as well as the spread of computer and Internet-based technologies, has aided in the emergence of the so-called "knowledge economy," in which true competitive advantage is found in information, knowledge, and human skills. These changes must be recognized and adapted to by tourism destinations. Tourism intelligence must be better organized, coordinated, analyzed, and shared among tourism actors in a knowledge economy to optimize the effectiveness of public actions at all stages. One area where public policy interventions can be beneficial is the spread of information resources to the tourism industry through better use of research-based intelligence (on issues like market trends, for example). Internet technologies have energized consumers by giving them unprecedented access to information about travel and tourism products, as well as the potential to remove traditional intermediaries, comparison shop, and create their own product packages. Tourism has been a pioneer in the use of information and communication technology to help visitors; Internet booking and marketing capabilities are progressing far beyond online booking to artificial intelligence.

The Internet has fundamentally altered the international tourism industry's interactions with customers. Firms of all sizes, from everywhere in the supply chain, can now connect directly with their customers. Also, whereas the consumer used to be at the bottom of the value chain, he or she is now at the top. The consumer now has direct access to the supply-side of the economy. As a result, the consumer now "governs" the tourism value chain. Consumers can identify their product choice in a variety of ways. They can switch from direct to indirect travel distribution channels, take advantage of relative online cost transparency, make informed comparisons, and sometimes even set their own price for the tourism product they desire. A key challenge now confronting public and private tourism actors is ensuring that the quality of information provided to the consumer is fully competitive with the consumer's next best option. The stakeholders who position themselves most effectively in the value chain will be the ones who benefit the most from the flow of information to and from the consumer, thereby improving their market position.

7.7.6. Human Resources

Labor market problems are critical to the growth of tourism. In developing countries, competition for labor from sectors of the economy with higher productivity growth, the opportunity to pay more, and better job development opportunities is causing a labor or skill shortage in tourism. This has increased the demand for immigrant labor. Bad perceptions of working conditions, career paths, and pay scales in comparison to other sectors are likely to persist. An aging population can present opportunities for reskilling in industries like tourism. The provision of adequate education and training schemes is a critical area for promoting creativity and improving productivity in the tourism industry.

In labor markets, the tourism industry faces a variety of obstacles, including: difficulties in hiring the best workers, as well as retaining and improving them over time. To keep up with evolving tourism trends, the industry is constantly in need of new skills. Failure to address labor or skill shortages effectively can stymie tourism development, as low quality may reduce productivity and harm the industry's competitiveness. Despite the current economic downturn, the long-term prospects for tourism growth are favorable, and the need to attract, train, and maintain high-quality labor is critical to the industry's long-term growth. There is a need within the education and training system to upgrade the skills of tourism-sector workers for a variety of reasons, including increasing industry productivity, equipping tourism-sector workers to respond to the realities of the knowledge economy, and ensuring that skills exist in areas such as sustainable tourism practice and the increasingly important area of green practices within the industry. The adoption of a philosophy of life-long learning is one field that needs to be prioritized in order to improve both ability levels in tourism and career growth opportunities, making it easier to retain key personnel. Education and skill development are critical for all tourism destination countries, but in developing countries, education, and skill development will present a significant challenge, especially in those countries experiencing the fastest tourism growth. Developed countries would be forced to adopt training and capability creation as a means of raising productivity if they are to sustain competitiveness in the face of increased competition from lower wage competitor countries and increased pressure on service quality standards. The predominance of SMEs in tourism service providers poses special challenges in terms of ensuring that training and skill development programs are accessible to smaller businesses.

As countries such as China and India take up an increasing share of international tourism flows, the development of new high growth markets causes changes in the cultural characteristics of established tourist trends will present specific demands for the production of suitable language and cultural skills. Attention must also be given to the problem of migrant labor, which is currently a significant source of tourism-sector employees in several destination countries but also puts unique demands on education and training providers.

7.7.7. Productivity and Competitiveness

Tourism industries in OECD economies face fierce price competition from competitors in developed countries, owing to the labor-intensive nature of tourism and the generally higher wage costs of OECD member countries. Tourism is also competing with high-growth goods and services that are becoming increasingly important in households' discretionary spending, such as mobile devices and ICT services. Many sectors in OECD economies have expanded faster than tourism in recent years. Tackling productivity is thus becoming increasingly important in the tourism industry, particularly to moderate price increases in the sector, allow for better labor conditions, and design new appealing products, attract new investment, and innovate.

Poor labor conditions exacerbate the industry's high labor turnover and weaken its ability to compete in labor markets. A lack of investment is also contributing to an aging tourism infrastructure in mature tourism destinations. To address these issues, both supply- and demand-side policies, such as value-based pricing, promoting entrepreneurship, developing efficient frameworks for innovation, making better use of IT-based networks, and ensuring that service quality remains high through appropriate human resource management practices, are needed. OECD nations, on the other hand, can have a range of advantages. OECD participants, for example, have a potential expertise advantage over developing-country competitors due to a trained population, access to technology, administrative, and business skills, stable, and competent political and administrative leadership, well-established banking and financial organizations, marketing skills and experience, and generally successful infrastructure. In addition, the expansion of broad global travel and tourism networks in OECD countries can have an effect. These benefit significantly from vertical, horizontal, and diagonal integration, as well as scale and scope economies, and can fund major investments in modern electronic databases and marketing. Such large travel and tourism operators are likely to accommodate to an increasing proportion of mainstream and mass tourism movements, accommodation, and related services.

At the same time, the expansion of large operators poses specific risks to SMEs, which comprise the majority of tourism industry enterprises and play a key role in smaller regional locations. The above-mentioned increasing fragmentation of demand, as well as the growth in demand for unique and specialized products and services, suggest that scope will remain for smaller and niche operators. Improved networks with other businesses, government agencies, and industry associations, as well as along value chains, could be solutions for SMEs to increase business efficiency and communication.

7.7.8. Essential Factors for Effective Tourism Policy Development

Tourism has a major economic impact in many countries and non-member economies. However, many countries find it difficult to concentrate on the tourism sector. The tourism sector is overseen by a number of ministers, including those in charge of the economy, trade, industry, regional development, culture, sport, and transportation. There are only a few countries that have a dedicated Ministry of Tourism. Why should this be the case? The primary reasons appear to be historical in nature, in that tourism was relatively late in recognizing its economic and developmental potential and importance, whereas other sectors, such as agriculture and industry, have long had dedicated ministerial portfolios to nurture and encourage them. As previously stated in this chapter, tourism crosses many ministerial portfolios, making it difficult to identify a single ministerial area of responsibility. Nonetheless, tourism has made significant progress in gaining government recognition as an economic activity worthy of serious consideration at the national policy level. Its economic contribution is increasingly being recognized as outstripping that of many other important sectors, such as agriculture, which has traditionally received far more political attention from governments at the national, regional, and local levels.

7.7.9. Planning Tourism for the Long-Term

As tourism progressively raises its profile in national economic planning, there is a clear need to ensure that attention is paid to its long-term development potential. Such an approach demands comprehensive strategic planning for the industry's future-partly for the reasons given earlier in this chapter, notably the need to maintain market share and competitive advantage by ensuring that the industry achieves productivity growth at least as successfully as other growth sectors. National governments need to take the lead in the industry's strategic planning. This process of developing a long-term strategy requires the adoption of a clear vision for tourism. It should be built on an open discussion with all stakeholders and should bear in mind the need to involve all relevant ministries to achieve a comprehensive cross-government response. The process of developing a tourism strategy generally involves several steps. These are illustrated in Figure 7.3.

7.7.10. A "Whole of Government" Approach

Industry priorities are complex and frequently uncoordinated in a highly competitive industry like tourism, were problems cross various ministerial roles and levels of government. The process of setting out and gaining a degree of consensus on the current situation faced by tourism and tourism destinations, the development of a longer-term vision, and the setting of goals, objectives, and measurable outcomes can be highly valuable in bringing key stakeholders together, sharing knowledge, ideas, and concerns, and helping to raise awareness generally within and across government and in the wider community. Traditional tourism industry strategies have tended to focus on the demand side of tourism and the need to market to the visitor in order to draw them to particular destinations. While this remains a valid component in strategic industry planning, an increasing number of strategies have been widened in order to include issues such as destination management and questions of industry "leadership" and direction. As a result, there have emerged new frameworks for policy action leading to policy objectives that include integration and collaboration across different agencies (i.e., "whole of government") and different levels of government, as well as partnerships with industry.

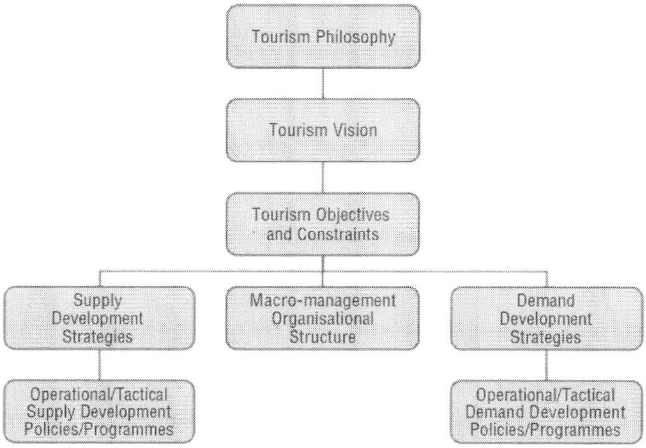

Figure 7.3. Developing a government strategy for tourism.

Source: Goeldner and Ritchie (2003).

Tourism planning strategies must progressively take into account the environmental concerns addressed earlier in this chapter, as well as consult with industry interests. As a result, the planning process entails several layers of government at national ministerial levels, as well as collaborations with industry and the private sector, whose members are also better positioned to identify places where government policy initiatives can be most helpful. As a result, tourism policies are increasingly reflecting the need for an integrated approach to tourism growth across a broad range of government ministries. A number of conditions must be met, as well as some important questions asked, for such an approach to be successful. For example, does the policy have a commitment from all of the appropriate authorities and levels of government to its aims and priorities and to the special interventions and measures proposed? Who is in charge of ensuring the implementation of policies that come under the purview of government departments other than tourism? How will the application and evaluation of these interventions be ensured? Clearly, the most beneficial outcome is for the policies to have authority in all related government ministries and to be coordinated at all levels of government. Global, state, and provincial tourism plans are often cabinet-approved documents. For example, through a government resolution in 2005, the Slovak government approved the Tourism Development Strategy of the Slovak Republic until 2013. Similarly, the Prime Minister unveiled the New Zealand Tourism Strategy (2015), emphasizing its all-government status. The New Zealand Tourism Strategy involves both expectations and annual success metrics, as well as a cabinet review mechanism. Collaboration and coherence at the federal, state, and local levels of government. It is also clear that the process of maintaining consistent and mutually beneficial policy structures vertically across different levels of government faces unique challenges in many member countries. This is emphasized where national government authority over state or provincial governments is restricted or opposing political parties' control each. Large variations in the scope of obligations and policy focus further complicate the situation. A national government may have little responsibility for land use planning challenges that are crucial at the local government level. Regional and local tourism organizations can see little value in high-level cross-government policy issues. The significance of this topic is well illustrated in Canada's document Building a National Tourism Strategy, which is subtitled "A Framework for Federal/Provincial/Territorial Collaboration." One of the next significant challenges for governments at all levels in enhancing the efficacy of tourism policies and planning appears to be the question of cooperation and continuity between levels of government in tourism strategic policymaking and implementation of particular programs and initiatives.

7.7.11. Industry Engagement

The significance of including the tourism industry in the strategic planning phase cannot be overstated. In several countries, there is an increasing commitment to the strategy as a collaboration between business and government. The New Zealand Tourism Strategy (2010), for example, was described as a tourism industry/government collaboration and was overseen and published by a Strategy Group comprised of influential private sector and tourism industry members. The Australian government announced the development of a National Long-Term Tourism Strategy in May 2008. This plan would have a long-term vision for the Australian tourism industry, with an emphasis on supply-side issues. To provide input on the implementation of the Strategy, a Steering Committee with representatives from both within and outside the industry has been formed.

7.7.12. Outcomes, Evaluation, and Performance Measurement

Best practice tourism plans have become more advanced in terms of establishing measurable targets and objectives and defining success metrics. An intensive process of engaging stakeholders in the production of the metrics and obtaining their adherence to agreed-upon performance measures is critical to achieving this. This entails defining performance metrics that represent the stakeholders' ability to influence outcomes and can be accurately assessed, as well as establishing agreed-upon and realistic time frames for implementation.

7.7.13. Policy Options to Promote Competitive and Long-Term Tourism Development

This section aims to highlight the areas where governments can make the most successful contributions, as well as to suggest longer-term tourism initiatives that would be suitable as the sector grows. Human resources, creativity, SMEs, destination accessibility, environment, and climate-related concerns, marketing, and branding, economic growth, culture, and local development, safety, and security, and measurement and assessment are the key areas where action is needed to resolve these challenges. Both of these problems help to maintain and improve the industry's competitive role.

Understanding how to improve the role of their tourism industries is a key policy question for OECD countries and non-member economies: the challenge is to stay competitive within a framework of sustainable growth. Since tourism industries in OECD member countries are relatively mature, it is important to refresh and expand the tourism product in order to retain a competitive position. Thus, one of the primary goals of government policy initiatives in the tourism sector is to sustain and improve the industry's competitiveness. But, in this sense, what does it mean to be competitive? Competitiveness in tourism refers to a variety of issues, including the country's position as a tourism destination in comparison to competitors, price, and value-for-money issues for visitors, productivity, and profitability within the industry, and the magnitude to which a given tourism sector is able to develop and refresh its goods and services in order to present high quality. Since tourism growth must align economic development with social, cultural, and environmental goals in order to ensure the sector's long-term viability, sustainability, and competitiveness go hand in hand. As a result, sustainability should be regarded as a significant competitiveness engine, especially in light of concerns about climate change and environmental protection. The growing value of green practices has implications for sustainability, and thus for the tourism industry's competitiveness. The OECD has discussed this topic broadly (in Green Growth: Overcoming the

Crisis and Beyond), a strategy that was formally adopted in the Declaration on Green Growth issued by an OECD Council meeting at the Ministerial level in June 2009. These sources suggest a large strategy for promoting low-carbon growth in both OECD member countries and non-member economies and announce an intention to promote this policy at the UN Framework Convention on Climate Change Conference in Copenhagen in December 2009. The numerous policy approaches listed-eliminating energy subsidies, removing trade barriers to environmental products and services, resolving market weaknesses, reforming policies with environmental goals, and actively encouraging green-collar employment-have significant consequences for tourism, but they also hold great promise. Tourism planners must be mindful of these goals and ensure that the tourism industry both contributes fully to their realization and positions itself to benefit most from policy reforms and green initiatives. Since tourism impacts and is affected by many other sectors of the economy, competition is a broad concept for policymakers at the global, regional, and local levels. Perhaps most importantly, the productivity through policy initiatives, including those linked to the green agenda, necessitates better integration of tourism and non-tourism policies. Tourism policies do not solve the broader challenges and constraints confronting the sector on their own, so improving the sector's competitiveness does not fall within the purview of a single ministry or policy directorate: it needs cross-cutting policies and coordination across the public and private sectors to produce effective policy outcomes for the tourism sector.

7.7.14. Human Resources Development

Tourism is a labor-intensive practice in which the tourist's experience of a given destination is heavily reliant on connections with local people and those employed in the tourism industry. Front-line workers must be educated and trained in customer service skills in order to provide high-quality tourism services to visitors. Most OECD economies report growing difficulty in meeting the demand for specific industry skills. Tourism plans should include long-term preparation for the tourism labor market, as well as education and training programs to address skill shortages and improve the efficiency and productivity of the industry's workforce. According to the OECD's regular publication, education at a Glance, the main focus in many OECD countries is on improving tertiary education opportunities for young people. However, such large commitments to expanding tertiary education are not inherently enhancing art- and skills-based training, which dominates the needs of the tourism sector. Along with an emphasis on higher education, many countries need investment and advancement in basic craft skills. Private hotel school programs assist in meeting training and management needs in some countries, such as Switzerland, but in many countries, the craft standard has been neglected to the detriment of the needs of the tourism labor force. This illustrates a major flaw in many educational policies, which can run counter to the tourism industry's need for entry-level workers who can be hired, educated, and retained by employers. Because of its comparatively low rewards and remuneration costs, the tourism industry continues to struggle to attract and retain employees on a global scale. Furthermore, the burden on public education and training budgets is causing many governments to consider alternative delivery models for the tourism sector in order to meet capability needs. To match business needs with private sector preparation, one choice now gaining policy acceptance is a move toward partial or full private sector delivery of training and manpower growth.

7.7.15. Boosting Innovation and the Knowledge-Economy

With increased global competition and relative declines in tourism growth rates in many OECD countries, innovation is now regarded as a major driving force for structural change and growth in the tourism industry. To stay competitive with new tourism destinations, conventional (OECD) tourism countries should be at the forefront of the creation of new products and ideas in tourism, as illustrated in the OECD publication on Innovation and Growth in Tourism. The main policy message of Innovation and Development in Tourism is that, in the absence of a proactive tourism policy, even the best tourism destinations will lose competitiveness and growth momentum. Tourism policies that are well-targeted will foster frameworks for innovation growth (e.g., human capital, knowledge, R&D, information technology), the production of new business models, and a dynamic entrepreneurial community. Public interventions must identify areas where innovation can occur in tourism services and assist in preventing market fragmentation. This is particularly important for small and medium-sized businesses.

7.7.16. Helping SMEs to Access Global Markets

SMEs play an important role in most countries' tourist experiences, but they face significant barriers to entering global markets. Travel and tourism SMEs now have more opportunities to enter foreign markets, thanks to the increased use of information and communication technologies (ICT) and lower transportation costs. Global Value Chains (GVCs) and networks enable SMEs to develop their know-how, creativity, and product quality on a continuous basis. Since the buyer is now at the center of the chain, SMEs will explicitly meet their needs. The difficulties, on the other hand, are various. Small and medium-sized enterprises (SMEs) must invest more in training and skill growth. They must adapt to new products and process requirements. Policy interventions are needed to promote this process and ensure that SMEs do not lose their ability to compete with larger suppliers.

7.7.17. Accessibility of the Destination

Tourist accessibility to and from destinations and origins is a critical necessity for the sector's continued development. The construction of airports in China, for example, with public and private sector investment and collaborations to increase airport capacity, exemplifies the interdependence of transportation and the tourism industry.

7.7.18. Addressing Environmental and Climate Change Issues

Tourism destination sustainability is more than a convenient term. It is critical to the long-term growth of the tourism industry as well as knowing the motivations of visitors to visit a specific destination. Many tourism plans mention sustainability and detailed strategies for ensuring that sustainability requirements are met. Furthermore, a greater commitment to green values on the part of the tourism industry is needed in this region. Energy consumption, the relevance of transportation sectors to tourism, and global concerns about climate change all provide opportunities for tourism policymakers to take constructive steps to help the tourism industry. Fuel price volatility for an industry that has traditionally evolved on the basis of cheap energy poses many policy issues for governments, especially in countries dependent on long-haul markets where energy cost decreases in new aircraft technology are unlikely to offset fuel price increases. According to the International Energy Agency, rising demand for transportation, a portion of which will come from visitors, will account for 20% of the increase in energy demand to 2030. Tourism remains overly reliant on oil as a source of energy, reinforcing the need for the tourism industry to consider alternative energy sources, especially green energy sources such as solar panels and wind power, to ensure its supply. OECD countries such as New Zealand and Australia, as well as many non-member economies that depend on long-distance origin markets, may need to reconsider their tourism policies and strategies. The relationship between climate change and tourism has emerged as a key policy issue. Although some positive effects may occur for specific destinations, climate change is widely regarded as one of the most serious threats to tourism (e.g., altered seasonality, heat stress for tourists from warmer temperatures, a wider geographical distribution of infectious disease vectors, the threat to small island states from sea level rises and the increasing frequency of extreme weather events and flooding). Policy interest has been devoted to the available solutions for mitigating the detrimental consequences of climate change.

7.7.19. Focusing on Marketing and Branding

The tourism strategies cover a broad range of proposals for assistance in the creation of new and existing market opportunities, whether geographical (e.g., China or India) or perceived niche markets. Marketing support, network encouragement, web-based marketing, technology, education, and training programs such as language training and service quality guidance, and the elimination of barriers to further development are all included in the strategies. This involves a focus on domestic tourism, a critical component of tourism that is often ignored in favor of foreign markets. Active marketing and destination branding are topics that are intertwined with the growth of new national and international tourism markets. Most tourism strategies in member economies place a strong emphasis on destination marketing and the creation of a national or regional brand. Indeed, for most countries, the marketing budget given to national tourism offices or their equivalents for foreign marketing purposes is the largest single item of public budgetary funding for tourism. Thus, tourism campaigns at all levels of government usually concentrate heavily on marketing concerns, with a particular emphasis on niche markets. Tourism strategies usually address destination branding and the need to compete with other destinations and industries-especially when addressing the creation of a national brand.

7.7.20. Promoting Economic Development

Tourism is generally accepted to be important for economic growth. Since tourists visit suppliers, the industry may play an important role in poverty alleviation by providing employment for unskilled or semi-skilled workers in hotels, resorts, and cultural sites, as well as promoting job growth in supply industries. Policies are needed to expect benefits are shared and distributed to vulnerable communities, as well as interventions to reduce the negative effects of tourism on the climate. Opening up markets for service trade and investment could have a significant impact on the growth of tourism. If sustainable tourism and development are to be achieved, any market opening must be carefully planned and enforced, taking into account the social and environmental effects, and integrating the new focus on green employment and greener business practices. Governments are attempting to resolve these issues through a number of policy responses. While information and data on the effect of reforms are scarce, there is preliminary evidence that reforms are resulting in significant gains for the tourism industry. More liberal air transport policies, especially open skies agreements, are increasing service capacity and lowering fares, allowing more people to travel to destinations. Innovative policies are also assisting in the improvement of physical infrastructure, which often lags behind tourism expansion. Partnerships between governments and investors, in particular, are enlisting the private sector to assist in financing infrastructure construction and operating the resulting services.

7.7.21. Valorizing Culture and Local Attributes

Tourism is an effective way of valorizing culture and generating revenue that can be used to sustain and enhance cultural heritage. Developing a good relationship between tourism and culture can thus help places become more appealing, as well as improve their attractiveness as places to live, visit, work, and invest. Governments can play a role in promoting cultural heritage and, as a result, increasing a destination's tourism appeal. A long-term perspective is especially important here, because improving a place's appearance or growing its appeal does not happen overnight. In most cases, it takes 20 to 25 years to reap the full benefits of long-term interventions in the fields of culture and tourism. Policies in this area that benefit residents would also be appealing to tourists, and vice versa. Many public organizations are involved in local tourism planning, growth, management, and marketing, but few have been explicitly developed to function in this area. Furthermore, cooperation among the different stakeholders is often not satisfactory. Territories also focus on competitive growth when they should also develop complementary tourism products. Setting up governance frameworks to increase tourism's competitiveness and efficiency at the local level, as well as ensuring the coherence and continuity of policy formulation and implementation, is thus a significant challenge. This will ensure long-term and healthy tourism growth, as well as strengthen the tourism industries and a destination's competitiveness at all levels.

7.7.22. Increasing Safety and Security

Traveler safety and security are important for the growth of the tourism industry (as for the global economy in general). Many international events in recent years have shown that tourism can be a focus of terrorism and global uncertainty around the world. Safety has emerged as a critical factor in traveler selection. Ensuring that tourists can fly and participate in tourism in a healthy and secure manner is now recognized as a vital determinant of a destination's competitiveness. To resolve this issue, governments are enacting new policies, such as those governing travel documents and entry/exit procedures, in order to help ensure the safety and security of both travelers and host countries. Tourism growth has been inextricably linked to safety and security policies. There is a need for greater international cooperation to harmonize various methods and share new practices in this field, such as when calculating the financial consequences of safety measures on travel and tourism. The main challenge for policymakers is to strike a balance between protection and security measures and freedom of movement.

7.7.23. Improving Measurement and Evaluation

Efficient national tourism organizations and sound market-oriented policies at the national, regional, and local levels contribute significantly to the provision and quality of tourist services. Effective tourism policies fill market holes, such as those in promotion, business networking, creativity, knowledge, and analysis, and education and training. They also contribute to the strengthening of harmony between tourism policies and other policies (for example, transportation or the environment) and to the sector's competitiveness. Evaluation is important for determining whether policies and programs are acceptable and effective in achieving their intended goals. Data from both quantitative and qualitative sources are used in the evaluation. While quantitative instruments are commonly used (often simple arrivals or expenditure-based measures), the effect of tourism on individuals and communities cannot always be reduced to quantitative measures; thus, qualitative tools must also be used to understand how policy initiatives could have affected tourism outcomes. Since measurement instruments and performance metrics are still in progress, measuring, and evaluating policy outcomes in tourism is still in its infancy. Although the TSA has received a lot of attention, it is still simply a static accounting tool for calculating the contribution of tourism to an economy. Even at the level of economic impacts alone, the TSA is unsuitable for calculating the effect of changes in tourism demand on key parameters (e.g., gross value added, employment) and thus for coping with many of the policy issues that governments face in relation to tourism. Other statistical advances in tourism research, such as Computable General Equilibrium (CGE) models, such as Scotland's Moffat Model or Australia's Sustainable Tourism Cooperative Research Center's Tourism CGE Model, aim to understand the effects of such changes and how they will impact both tourism and the destination's overall economy. Such models are now commonly used in a number of OECD countries, though their use in a tourism policy sense, in place of more restricted tools including input-output analysis, is still in its early stages. The following methods are commonly used for tourism policy analysis: cost-benefit analysis, input-output based assessments and target evaluation, as well as public engagement and feedback on policies, as shown by the public consultation process that is frequently correlated with new policies and strategies.

Chapter 8

Current and Future Perspective of Sport Tourism

CONTENTS

8.1. INTRODUCTION

Tourism is always changing, but there is a significant shift in demand and tourist expectations right now. Sport tourism brings together the best and worst of both the sports and tourism industries. Sport is a universal language that involves competition, conflict, emotion, and frequently entertaining drama. Tourism is inherently intrusive, involving host-guest interactions and impacts. Sport tourism typically brings money, pride, and entertainment to a city, but it is not always about happy tourists and their spending. It can sometimes result in crowding, queuing, environmental degradation, price gouging, and local resentment. According to sports trade analysts, this multi-trillion-dollar industry will see a significant increase in the coming years. The global developments in sport and sports tourism will be shown first, based on the premise that social development is always a question of creative power and the influence of stakeholders. It is clear that the western, and increasingly commercialized, sport model is being exported globally. In the second step, current trends in sport tourism will be presented.

8.2. FUTURE TRENDS IN SPORT TOURISM

Contributions to the future of social development are fraught with risk. Nobody knows what will happen in the future. These include, but are not limited to, sports organizations, the media, and local to international tourism politics, commercial sports providers, and tour operators. Above all, it is the tourists, who have no establishment or organization but could actively participate in the future development. Certainly, the requirements of the aforementioned stakeholders must be considered. Nonetheless, sport-related offerings, goods, and services are not a carbon copy of existing wants and needs. Particularly since these do not always need to be conscious, but can be implicitly present and thus more or less adaptable. Sport's development is becoming increasingly focused on the realization of the main stakeholders' interests. Golf media promotion in Europe is just one example among many. Figure 8.1 depicts the primary stakeholders in the development of the sport. The stakeholder groups with the least influence on the development of sports are depicted on the left. Stakeholder groups with a dominant position of influence are on the right side.

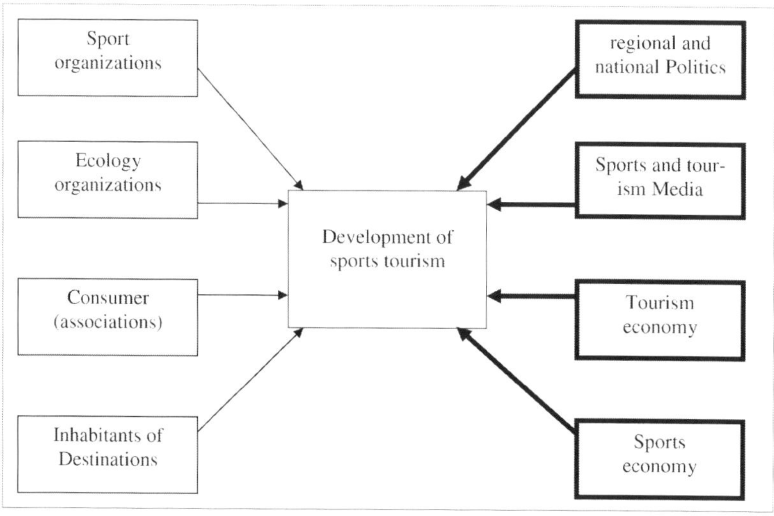

Figure 8.1. Social protagonists and sport development.

Source: https://www.researchgate.net/publication/248938711_Future_trends_in_sport_tourism_-_A_question_of_the_development_by_social_protagonists/link/5c4889dfa6fdccd6b5c2e7fc/download.

Not all stakeholders have equal power to influence developments in their preferred direction. The tourism industry, the sport industry, the sport-tourism media, as well as political stakeholders have the most influence on development. Following them are environmental organizations, consumer groups, sports organizations, and local residents. However, there are no consistent strategies among the various stakeholder groups. Competition among them results in either authority and leadership disputes or the formation of strategic alliances. The ski industry in the Alps is a good example of this, where increasing competition leads to mergers in order to secure a better market position.

8.2.1. The Global Benchmark

Before we can label current trends, we should look at global developments as they have so far been demonstrated in sports and thus sports tourism, and will most likely continue in the future. What types of cultural transfer and social changes will occur as a result of sports tourism? From the diversity of existing component sport cultures, only those sectors will be exported, which the media, sports article and tourism industries, sports associations, and state 'development aid' either regard as marketable or are overlaid with central esteem. Torsion (2002) stated that there is no "equal distribution of worldwide relationships" in sports, but instead a specific western dominance. "Prevalent values, goods, and cultural patterns frame the character of the modern spirit, particularly in Europe, North America, and parts of Asia." Performance, individualization (not individuality), speed, hedonism, intoxication, and self-interest are examples of predominantly Western concepts and values that are exported (Figure 8.2).

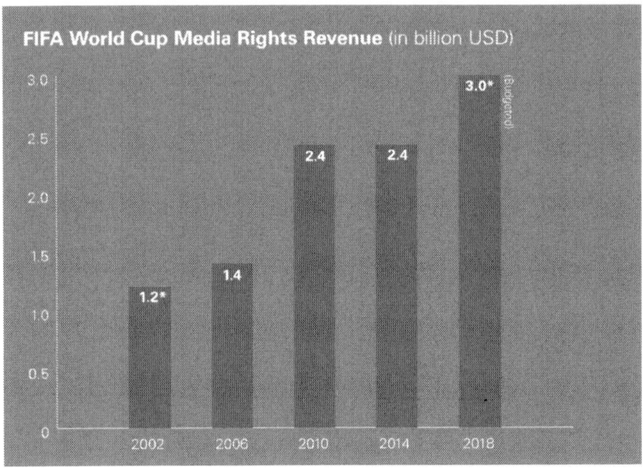

Figure 8.2. Current trends and global developments in sport tourism.

Source: https://www.footballbenchmark.com/documents/images/FIFA%20World%20Cup%20Media%20Rights%20Revenue.jpg

Furthermore, East Asian exercise habits are transformed into a practical type updated for western purposes, in which their metaphysical heritage is often stripped away, replaced by an esoterically oriental philosophy, and then re-exported via tourist offers. Influences from the alleged peripheries can be found in sectors of east-Asian exercise practices, as well as varied dances from various continents, including Polynesia. The alleged center adapts and culturally changes these influences (as the west sees itself). Surfing evolved from wave-riding, and tower-jumping evolved into bungee-jumping. These last two-sport forms are being re-exported globally, but with a western sense and importance culture. The export of the sport culture that has developed within this center, however, has remained dominant. Football, golf, tennis, and motorsports are the standard-bearers of this global sports culture (Figure 8.3).

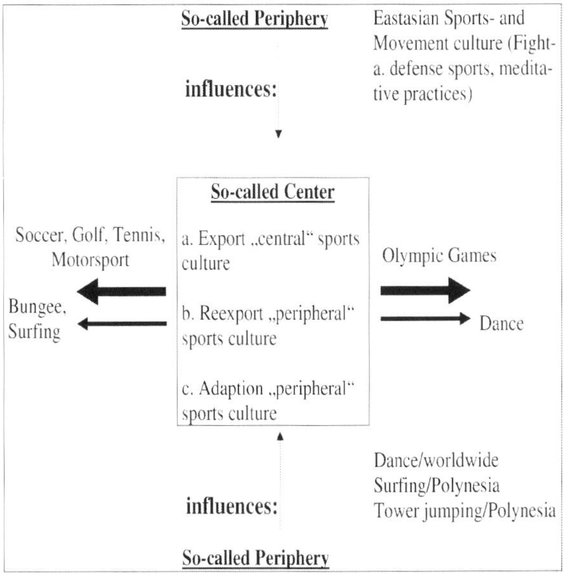

Figure 8.3. Analytical model of an asymmetrical sports cultural exchange.

Source: https://www.researchgate.net/publication/248938711_Future_trends_in_sport_tourism_-_A_question_of_the_development_by_social_protagonists/link/5c4889dfa6fdccd6b5c2e7fc/download.

In comparison to colonialization's mechanisms of sovereignty and oppression, globalization's cultural transition processes are more nuanced and adaptable. The culture industry's medial and goods-formed distribution and consumer politics do not proceed on the basis of a deliberate and direct degradation of indigenous values and norms. In the vast majority of cases, it is reasonable to assume that this was not the intention. Nonetheless, the effects and outcomes of globalization may be linked to those of colonialism. Cultural globalization will thus be, on the one hand, a particular neocolonial type based on a harmonized and harmonizing symbolic code that is transported through ideological medial networks into which subjects seem to voluntarily integrate. On the other hand, globalization's technological and informational moments provide enhanced local opportunities, regional self-reflection, cultural exchange, and their own further growth. Despite the classification of both sides of the coin, little is said about the true relationships. The billions of dollars spent on ads, which is increasingly charged with entertainment, as well as the oligopolistic media systems.

The negative and positive processes of the development of asymmetric factors allow for the recognition of different trends in both the historical and current contexts. The complexity in establishing a clear division between the culture, tourism, and sports industries, as well as the media industry, exists here, particularly because the processes execute a slow and not always openly noticeable growth, and exhibit class-specific differences: Loss of traditional sports culture; part-modification of 'newly' introduced sports types via traditional meaning and value patterns; 'new' sports types as a foundation for broader development; return, conservation, as well as preservation of traditional sports types (dances); and finally modernization and development of traditional sports types via cultural influences.

8.3. CURRENT TRENDS IN SPORT TOURISM

8.3.1. Commercialization

With the development of so-called everyday cultural sports systems, and also some service sports systems, it is increasingly commercialized sports providers, in closer relationship with the sports article industry and the media, that have an ever-increasing influence on sports and sports tourism trends. Sports organizations, with their teams and alliances, are losing importance and decision-making authority in this regard.

8.3.2. Differentiation and Multioptionalism

Diverse sub-disciplines emerge and distinguish themselves from individual sport types as a result of the market's practical logic, or new sport types are invented or adopted. For each of these sub-disciplines, the sports article industry creates new equipment and clothes. Walking and Nordic-Walking was added to the list of activities that can be done in addition to jogging. Free climbing is a sport that complements climbing. Kite-surfing has entered the ranks of surfing. The list can be extended at any time and will almost certainly continue to develop in the future. In comparison to the earlier team or club-oriented sports model that young middle-class male competitive athletes were familiar with, today's athletes do not perform only one sport style, but rather a variety of option (sport hopping). In the summer, for example, windsurfing, and mountain biking, and in the winter, snowboarding. Diverse excursions are often included, where various sports styles are tested as the opportunity arises.

8.3.3. "Anything-Goes-Principle"

Offshoots of the differentiations, on the one hand, contribute to a so-called "de-sporting of the sport." Animation events that have been stripped of all sporting significance achieve the label 'sport' as a result of superior marketing. The variety of these deals ranges from 'Banana-BoatRiding' to 'Tubing' and 'Zorbing.' These superficial, artificial experiences are based on sensationalism and superlativism. The emergence of a hyper-reality.

8.3.4. Target-Group Orientation

A further increase in sport tourism can also be expected in the future. Above all, we should consider an older demographic group (aged 40 to 60) that has been able to incorporate sport as a more regular part of their lifestyle. In the 1950s and 1960s, a sports career after the age of 35 was scarcely viable due to the exclusively achievement-oriented, team-based sport structures. Wide-ranging sport experience, sport-related post-secondary education options, and technological advancements in sports equipment that make for an easier introduction (e.g., Carving-ski) contribute to daily participation in sports well into the 60s. Only after that age does daily participation in sports activities decline. Women's involvement in sports has also increased significantly.

8.3.5. Qualitative Offerings

More and more sports participants have a wide range of sports experiences and skills. As a result, there are qualitatively high pressures on sports facilities and sports education organizations. The certification of vocational and avocational sports bodies achieves a higher level of importance, as shown by stronger partnerships between tour operators and universities, sports associations, and other interested parties. Overall, there will be a greater demand for sports coaching, which must be tailored to the individual. In order to ensure professional consistency both internally and externally, it would be ideal to establish mutually agreed certification criteria, as commercial sports providers (Association of Fitness Clubs) have already done. Above all, a widespread confusion within the service and tourism sectors must be mentioned. The preparatory technological and organizational elements of sports offerings with so-called expertise and/or adventure characteristics are often displaced. This can have a completely rational purpose, on the one hand to increase the flow of tourists, and on the other to obstruct the well-intended service aspect, including advancement, approach, clothing, knot-making, and securing, with which the tourist is typically obviously burdened.

According to Gerhard (1993), experiences occur only after a situation and a subject interact. This means that the subject conducts an integration performance in response to the external event. Furthermore, in order to maintain their grip on the experience, the subjects respond to it by reflexes. As a result of this, perceptions cannot work by themselves. There is no such thing as a guaranteed pledge of experience. It is up to the person to make his or her own decision. It is to be questioned, however, on the basis of which expertise and knowledge base they make this decision, and what contribution sports tourism-related stakeholders can make to increase this experience and knowledge capacity.

8.3.6. Self-Reflection and Boundary Experiencing

As an expression of individualization tendencies, contracts which can attribute self-reflection and/or boundary experiences are being urgently needed. These offers come from the field of east-Asian exercise activities (in the broadest sense of the term) on the one side, and adventure and experience offer on the other. The need for identity, self-reflection, as well as self and boundary experiencing, has increased with the demise of concrete, physical work-related operation, increasing physical forces such as overall from acceleration and compression processes. On the one hand, so-called integrated offers with heavy influences from the East Asian domain, and often also with esoterically underlying extensions, cater for bodily and physical experiences. The vast majority of so-called adventure and experience providers were established in the 1980s (central value: adventure) and 1990s (central value: experience). The offerings from the evaluated providers range from soft offers with event and enjoyable characteristics to survival and intense tours. Individual boundary testing and self-validation will continue to be a growing field of (sports) tourism.

8.3.7. Nature and Contemplation

In comparison to the fast-paced fashion model, demand for sports activities will continue to evolve in a largely primal manner, in relation to criteria such as peace and needs, as well as regional identity. Because of its opposition to mass-tourism business strategies and the realities of the types of offers' limitations, this sector will be given a luxury character, irrespective of its latently ascetic orientation. We are dealing with an obvious inconsistency here. How can an ascetic way of life be luxurious? From an economic standpoint, we are concerned with stability, honesty, and nature as so-called scarce resources, which are often constantly threatened by being sold off or locked. With the caveat that the structure of daily life cannot meet the aforementioned requirements, or that external forces make it increasingly difficult, the growing importance of this sector is apparent.

8.3.8. The Demand Side

Regular participation in sports and cross-border travel are still mainly reserved for the G8, the wealthy and middle classes, and therefore a minority in the western developed world (Table 8.1).

Table 8.1. International Tourist Arrivals by Region (Millions)

	Base Year 1995	**Forecasts 2010**	**Forecasts 2020**	**Av. An. Growth Rate (%, 1995–2020)**	**Share (%, 1995)**	**Share (%, 2020)**
Africa	20.2	47.0	77.3	5.5	3.6	5.0
America	108.9	190.4	282.3	3.9	19.3	18.1
East Asia/ Pacific	81.4	195.2	397.2	6.5	14.4	25.4
Europe	338.4	527.3	717.0	3.0	59.8	45.9
Middle East	12.4	35.9	68.5	7.1	2.2	4.4
South Asia	4.2	10.6	18.8	6.2	0.7	1.2
Total	**565.5**	**1,006.4**	**1,561.1**	**32.2**	**100**	**100**

Within the sports types structure, distinctions emerge based on the variables of men/women, old/young, as well as the standard class variables of education, vocation, and income. Participation in sports on a regular basis remains an exception, for example, in the Federal Republic of Germany. According to reliable statistics, nearly 30% of the adult population participates in athletic activities once or more days a week. In recent years, the number has continued to rise. The majority of publications, however, indicate a significantly higher figure of participants due to the methodical nature of a holiday for many holidaymakers. The values are slightly higher here. To be sure, this is an intentional inclusion of irregular competitors in sport. Just a small percentage of holidaymakers list sports as a primary goal while planning a trip. Sport is only mentioned by 6% to 8% of holidaymakers as the most important activity on their trip. In recent years, this value has remained constant and has not increased. Nonetheless, sport is one of the many items for which calculation is not feasible due to the question's methodically difficult structure. Swimming (bathing) and walking are examples of movement behaviors that do not always qualify as sports (Table 8.2).

Table 8.2. Sport as a Main Activity in Vacation

	1995	**2000**
Very Important	8.2%	7.7%
Important	14.0%	14.9%

Although sport has been a stable cultural activity for decades, demographic factors (at least in the Western industrialized nations) due to changing age structures of society will cause problems for the tourism industry as a whole, and especially the sports tourism sector. Overall, the global population is declining, although the proportion of people over the age of 60 is increasing. The extent of the reforms would result in little to no improvement until 2010. According to the Willy-Scharnow-Institute of the University of Berlin's projections, population figures will be reduced by 8% by 2030, and by 20% by 2050. This would have a significant negative impact on the number of tourists.

8.3.9. Destinations and Climate Changes

When it comes to the potential of sports tourism, there are two trends that cannot be overlooked. On the one side, there is global warming, which would have some existential implications for the ski industry. On the other hand, the growing water scarcity that is currently affecting the Mediterranean region would, at the very least, have an effect on parts of Africa, south of the equator in the coming decades. The situation is exacerbated by rapid population growth. The local population in these areas is expected to double or triple, according to estimates. In the middle of the next century, every 4th individual on the planet will face a water scarcity.

Water-intensive activities, such as golf, would be particularly affected. Kenya, Morocco, and South Africa are three tourism-important countries that, according to experts, will join the list of countries experiencing water scarcity over the next 10 years. Global warming endangers alpine winter sports, and even more so low-mountain winter sports. According to current climate model forecasts ECHAM (European Community and Hamburg Atmospheric Model) and CCC (Canadian Climate Center), only ski resorts above 1600 to 2000 meters above sea level would be able to depend on a decent long-term snow cover in 30 to 50 years (Figure 8.4).

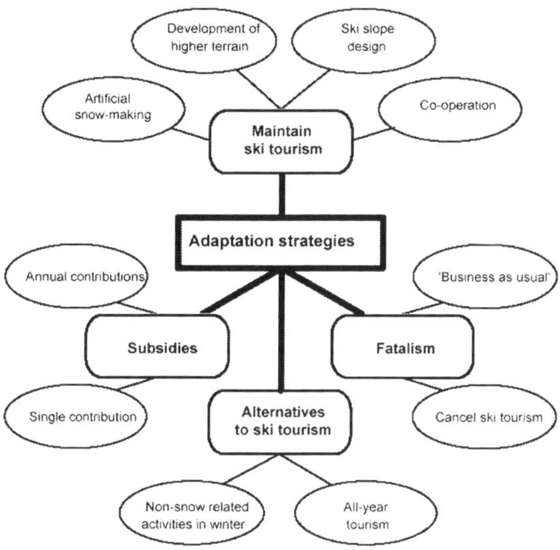

Figure 8.4. Adaptation strategies for snow uncertainties.

Source: https://www.researchgate.net/publication/248938711_Future_trends_in_sport_tourism_-_A_question_of_the_development_by_social_protagonists/link/5c4889dfa6fdccd6b5c2e7fc/download.

8.4. E-SPORT EVENTS WITHIN TOURISM PARADIGM: A CONCEPTUAL FRAMEWORK

E-sports (electronic sports) are organized video game competitions or tournaments that culminate in regional or international championships. The e-sports industry is rapidly expanding. The industry was projected to produce more than US $1.48 billion in sales by 2020. The number of viewers who track the industry on a daily basis and tuning in to watch international competitions was expected to hit 589 million by 2020. There are several reasons why esports events are appealing and memorable experiences: fans get a chance to (a) watch the best players in the world compete; (b) meet their favorite teams; and (c) share their excitement with hundreds of other fans. E-sports tourism, a hip version of event tourism, has been flying under the radar for a long time but has gradually reached the mainstream in recent years. This section here elaborates esports activities inside the tourism paradigm to examine the relationship between e-sports and tourism. As a result, e-sports activities, as a growing market, are a significant draw for the tourism industry.

8.4.1. E-Sports and E-Sport Events

Tourism has been evolving ontologically for decades. Socio-structural, socio-cultural, demographic, economic, and technological changes have often driven changes in tourist demand and forced industry suppliers to adapt significantly. With this in mind, many scholars have identified people's desire for quality of life and escape from the "pluralization of lifeworlds" and the "rationalization of contemporary urban life" as major motivators for travel (Rojek and Urry, 1997; Giddens, 1999). Tourism suppliers have concentrated their efforts on more homogeneous groups of visitors at the "hard or specialized" end of the market, referred to as "serious leisure participants," such as those pursuing adventure tourism, event tourism, eco-tourism, sports tourism, and cultural tourism (Trauer, 2006). Tourism consumption and production trends, as seen in the growth of "special interest tourism," have come to represent the growing variety of interests sought in late modern leisure society. In other words, a bilateral trade between tourism demand and supply has begun to affect the movement and development of new tourism experiences, as well as access to them (Douglas, Douglas, and Derret, 2001). Since the 2000s, internet-driven tourism has grown in popularity as a result of online communities focused on a particular or special interest, such as online gaming or modding,' all over the world (Cow and Young, 2013). At the same time, Getz invented the word "event studies," which he used in passing in his speech at the Events Beyond 2000 (Sydney) conference (Getz, 2008). As a result of these advancements in tourism and technology, competitive computer gaming, or "esport," emerged as a phenomenon, became a fundamental component of digital youth culture (Funk, Pizzo, and Baker, 2017), and has now begun to influence the tourism industry as a new travel motivation (Wagner, 2006; Yenişehirliolu, Erdoan, Ahin, and Ulama, 2018). Competitive gaming is referred to as e-sports, or 'electronic sports.' Young men and women from all over the world make a living by actively playing video games, typically by prize pools in tournaments or corporate sponsorships (McTee, 2014).

The market practices of content owners/providers (game developers) are included on the supply side of this phenomenon.

8.4.2. Relationship Between Sport, Tourism, and E-Sport

Sport, in its various forms, has always played and continues to play an important role in all cultures (Guttmann, 2004). Sport is an important component of tourism, and tourism is a fundamental feature of sport (Hinch and Higham, 2001); the definitions of tourism and sport are strongly linked and overlap. Sport-related tourism has grown in importance in recent years, both as an academic field of study and as a common tourism product (Gibson, 1998). Delpy (1998) defines sports tourism as "traveling away from home to play sport, watch sport, or visit a sports attraction," which includes both competitive and non-competitive activities. According to Delpy (1998), sports tourism can be divided into five major categories: attractions, resorts, cruises, tours, and events. Each of these categories is linked to other tourism sectors like adventure tourism, leisure tourism, health tourism, nature tourism, educational tourism, and virtual tourism. In recent years, a new athletic activity has arisen and gained prominence among young people: e-sports, which challenges conventional sports in several ways (Jonasson and Thiborg, 2010). E-Sport organizations like the world cyber games (WCG), Electronic Sports World Cup (ESWC), and Cyberathlete Professional League (CPL) organize LAN tournaments at both the national and international levels in collaboration with companies in the computer game industry (Hutchins, 2008; Jonasson and Thiborg, 2010). There are many explanations why esports competitions are appealing. Fans have the opportunity to watch the best players in the world, meet their favorite players, and share their enthusiasm with them.

Fans are able to drive thousands of miles to attend big events for these and other purposes. As a result, there is the potential for a new form of tourism to emerge: esports tourism. Indeed, according to Agius (2015), the idea of e-sports as a niche tourism attraction is viable, and with the considerable interest already shown in this market, it has the potential to become quite profitable. Other scholars, on the other hand, have noted that e-sports are a subset of leisure and that e-sports tourism has already grown in modern society (Seo, 2016; Lokhman, Karashchuk, and Kornilova, 2018). E-sports is officially defined as "a type of sport in which people develop and exercise mental or physical abilities with the use of information and communication technologies (ICT)" (Wagner, 2007; Seo, 2013). In other words, unlike conventional sports such as hockey, baseball, and soccer, e-sports require the interconnection of various platforms (Ayar, 2018). E-sports, also known as competitive gaming, are organized online video game tournaments that have grown in popularity as both a professional sport and a medium of entertainment (Heere, 2018). "Esports remains an industry with tremendous potential for the years ahead," says Wouter Sleijffers, CEO of Fnatic (Esreality, 2016). E-sports is a very complex space, and it is critical for existing and global esports brands like Fnatic to track developments using accurate and complete data. The market research company Newzoo released the first comprehensive study on the global e-sports market in 2015, estimating that e-sports generate US $612 million and have 134 million viewers worldwide. According to an overview of the most recent data from 2015, revenue in 2017 was projected to be US $325 million worldwide. The e-sports economy will rise to US $463 million in 2019, a 43% increase year on year, and will entertain an audience of 131 million e-sports fans and another 125 million casual viewers who mostly tune in for major international events (Newzoo, 2017). According to Newzoo, the industry was projected of worth $1.5 billion by 2020. (Tan, 2018). To summarize, e-sports and e-sports activities have seen phenomenal growth in recent years and continue to expand on an annual basis.

8.5. FUTURE PERSPECTIVES OF TOURISM INDUSTRY

When it comes to the future of sport tourism and its possible impacts on lifestyles, there are some changes on the horizon, especially in terms of sport tourism happenings and expectations. People can simulate games and events using Virtual Reality, and with advanced technology, the Internet can display every move of every event in minute detail. With interactive TV, the Internet, and imaging available, as well as the latest 'high speed chips,' the ability to perform complex tasks at nano-second speed, TV is increasingly integrating with all modes of communication, including mobile communication formats. The possibilities for sport tourism are limitless, as today's games will become tomorrow's realities for sporting fun, happiness, and fulfillment.

8.5.1. Sport Tourism Industry

Will sport tourists want to travel to various events when sounds, sights, and atmosphere can be produced digitally and surrealistically? Will there be a temptation to physically displace oneself when events can be modulated, interactive, and enjoyed in the comforts of one's own home or local establishments through current and future technologies? The virtuality of technology has also had an effect on the simulation of learning sport skills. This concept can also refer to a variety of sports tourism attractions, some of which can be found in museums and halls of fame dedicated to sports tourism. Individual satisfaction is increased by the immersive and supervised environments of the sports tourism experience.

Similarly, magnetic polarization will stimulate sports tourists during their resort visits. In this case, virtually has an effect on skill concentration and growth. Sports tourists are showing an increased interest in Sport Tourism Adventure, which can also be virtual reality. Above all, attendance at sporting events could rise. However, technological advancements have the potential to transform entire operating processes and structures, not to mention their theory. Natural terrain attractions such as trekking and mountain climbing in Nepal, tubing in the Grand Canyon, and white-water rafting are becoming more appealing and enjoyed by clients all over the world. In essence, various, and diverse societies are gradually recognizing the factors that contribute to sporting adventures. Furthermore, the cruise industry has acted as a catalyst for progress in the sport tourism sector. Sport tourism considerations of note include 'Cities on the Sea,' with ships the size of football fields offering condominiums and state-of-the-art sport facilities, as well as customized programs and services.

8.5.1.1. Potential Growth

Sport tourism, an increasing segment of tourism, has been shown to account for 32% of global tourism receipts. There is no question that much touristic activity has an impact on sport tourism, opening up many new possibilities and developments to promote entrepreneurship. Furthermore, lifestyle job trends all over the world are invariably contributing to the growth of sport tourism and its value/benefit impacts. Here are some examples:

- The short-term vacation opportunity is proving to be a beneficial factor of the tourism industry, reacting well to the pursuits of many attractive sports tourism destination experiences.
- Travelers who are inspired and committed to the activity in question can take a long-term vacation, resulting in a greater economic effect on the chosen sports tourism destination as well as a loyalty benefit to the activity in question.

- Sport tourism "destination dabblers" are busy, hardworking people who want to get away from the pressures of their careers. They are able to pay for short trips to participate in their desired events. These sports visitors can be seen tubing in mountain valleys, sea kayaking, and scuba diving the reefs, to name a few activities.
- Another factor involves sports visitors and their families, who can be seen visiting and studying historical sport viewpoints, eager to be satisfied with education, history, environmental awareness, and understanding.

8.5.1.2. Innovative Experiences

Diverse advances in sport tourism are assisting and leading to the expansion of offerings. Indoor simulation on a wide scale of natural and/or artificial outdoor facilities is sustaining sport tourism activities of fun and unique events inspiring the diversities within evolving profiles of sport tourists. Cruise ship ice rinks and virtual reality luge rides are two examples of simulation. There's also skydome colored snow and dome golfing, as well as gravity-free sports. A good example of value marketing and demonstration can be seen in the Key West region, where over $100,000 was spent on bait during a three-day mackerel fishing tournament. Fishermen's expenses on transport, lodging, food, and tackle equipment resulted in economic impacts.

8.5.1.3. Particular Developments

The use of natural surroundings promotes the development of resorts in smaller urban areas, as illustrated below:

- The use of specific facilities (skiing, scuba diving, golfing) has prompted the establishment of specific resorts, which have become the primary economic stimuli. Lake Placid has used the natural terrain of the Adirondack Mountains to host two Winter Olympics and to create a number of ski resorts in the region (Research Unit, 1998).
- Snowboarding on mountainous terrain is assisting certain ski resorts in their growth as more families participate in this activity. Data from some ski areas, for example, indicate that there are more than 150 new snowboarders on weekends, most of whom are children.
- Mountain biking clubs organize competitions in BMX, road

racing, motorbike trail and track racing. These sport tourists are fitness-minded individuals who transition from mountain biking to road racing and vice versa.

- Sport events in the wilds of nature are being more completely controlled, administered, and packaged. These experiences were designed and structured to value and benefit specific types of tourists. Gaming in Kenya and Sportings in Gambia are two examples of such management, which have resulted in new types of sport tourism.

8.5.1.4. Group Focus

Meetings, incentives, conventions, exhibits, shows, and fairs (MICES) have also proven to add value and benefits to sport tourism. Here are some examples:

- Meeting planners and organizers incorporate sporting activities to increase attendee retention.
- Incentive travel in the context of sport tourism is on the rise. This rise is specifically attributed to the worldwide popularity of participation sports and widely visible mega-events.
- Each MICES activity has an attraction that draws visitors to host cities and is considered a "shoulder" or "low" touristic season activity, as well as a "support" mechanism for the sports tourism industry.

8.5.2. Specific Components Classifications

Six distinct component classifications were proposed in order to forecast potential characteristics and benefits of sport tourism:

1. ***Social Value Component:*** Contributions to sport tourism in this category are intangible. Individual sport tourism perceptions are influenced by cultural appreciations, the status of the practice in progress, as well as travel forecasts and realities. The degree of social acceptance and satisfaction varies from person to person and group to group. Clearly, transparency and transferability are heavily influenced by demographics.
2. ***Interpersonal Value Component:*** The future of sport tourism has an effect on individuals both directly and indirectly. Participation, whether active or passive, may have an impact on one's actions,

character, value philosophy, and ideology. Evidently, opposing, or negative elements may be produced, which can have a negative impact on one's emotions and spiritual appreciations. Lifestyle interactions, psychological preparation, and multidimensional meanings are all applied to these interpersonal relationships.

3. ***Social/Knowledge Value Component:*** Without a doubt, some level of learning occurs anywhere sport tourism is depicted, whether professional or amateur participation. Self-evaluation is important for growth and maturity. Sporting events aid in skill activation and knowledge rating within the confines of a specific operation. The more one understands these trends, whether as a participant or as an observer, the more appreciative of the demonstrated talents and activity composition one is likely to become. In other words, simply being associated with a particular sporting activity has the potential for educational benefit, with the potential for incorporation into one's skill habits and decisions.
4. ***Health Value Component:*** Participation in physical activity demonstrates fitness or bodily wellbeing. Imitation and emulation are both direct and covertly encouraged in order to improve one's lifestyle. These positive principles develop, resulting in proper diet, physical activity, healthy sleeping habits, and active attention to one's essential bodily functions. This value aspect, as it develops in the future of sport tourism, has the potential to have a significant effect on work habits, family relationships, and societal policies. In general, sport tourism benefits an individual's mind and spirit, which can help to foster a better outlook on one's environment and lifestyle.
5. ***Marketing/Publicity Value Component:*** The spine of viable communities in the promotion and sale of a country's region, culture, traditions, and more. The more people worldwide who are familiar with and understand a specific tourism destination, the more appealing the destination is likely to become. How far-reaching such a value can be considered is dependent on the respective sport offerings and services. Without a doubt, the higher the level of complexity of the experience, the higher the inducement and retention factors. The magnetic pull begins with the very first imaginative projection and future expectation. Sport tourism affects and will impact potential viability and diversity in an unusual way.

6. ***Economy Value Component:*** Putting on a sport tourism show or event can be expensive in terms of finances, volunteer efforts, and community cohesion. Specific facilities may need to bc upgraded; state-of-the-art systems and equipment may be required, roads may need to be repaired, and infrastructures may need to be modernized. Diverse funding sources can become essential. These economic factors trigger a mini-boom, which typically maximizes benefits in the short term while retaining value for a longer period. Communities thrive when proper preparation is implemented, and visitors feel pleasure, happiness, and fulfillment. And, in the grand scheme of things, the world becomes smaller, and those who live in it may become more compassionate and knowledgeable of one another.

8.5.3. Evolutionary Trends in Sport Tourism

Sport and tourism have a long-term relationship that is just as complex as it is in the short and medium-term. Sport influences the life cycles of destinations, and tourism influences the life cycles of sports. Nostalgia s Sport tourism, which is inspired by a desire to reconnect with one's sporting history, is an example of how sport heritage can be used as a tourist attraction. Sport halls of fame, locations of past sporting events, and imaginary pasts carried out by fantasy sport camps and programs are some examples. The relationship between life cycles and nostalgia sport tourism does not occur in isolation from other influences at work in the larger world. These phenomena are influenced by a diverse system of local, national, and global patterns.

8.5.4. Cyclical Relationships in Sport and Tourism

Tourism's dominant characteristics are destination and product life cycles. With its six stages of involvement, development, exploration, consolidation, stagnation, and either rejuvenation or decline, Butler's (1980) tourist area life cycle model epitomizes this concept. In revisiting Butler's model, Johnston proposed that the early part of this cycle could be classified as the pre-tourism age, in which another institutional framework governs the destination rather than tourism. Similarly, the latter stages of stagnation and decline can be characterized as a post-tourism period, in which tourism has succeeded as a dominant or significant sector of the destination economy. These cycles follow a general trend of increasing visitation until the destination's resources are depleted, at which point visitor numbers begin to decrease. These cycles have a number of management consequences, the most obvious of which is that management action is required to maintain tourism capital so that the destination's life span can be extended.

8.5.5. The Evolutionary Dynamics of Sport

Sport is constantly being used as the focal point of revitalization plans for tourism destinations. The successful implementation of this strategy necessitates that tourism managers be knowledgeable about the sport life cycle.

Individual sports, sports disciplines, and sporting events, including tourism products, have their own life cycles. They, too, go 'out of style.' And they are constantly having to contend with other leisure activities and events. In sport, too, there is a persistent need for individual sports and competitions to be adapted to the evolving needs of athletes and spectators alike.

Much as the number of tourists to a destination is a primary indicator of the life cycle of a tourist destination, the number of participants and spectators involved is a primary indicator of the life cycle of a sport. Other indicators of a sport's standing during its life cycle include the sophistication of its rule systems, the degree of skill growth and physical fitness, and, significantly, the magnitude of commodification and professionalization.

The slow growth of many highly organized team sports in recent years, as well as the rise of individualized and extreme sports nations, have clearly demonstrated the dynamics of sport.

Organizational sports participation is declining, as are the proving grounds from which top-level sports historically draw new blood. The new generation moves in a sliding, gliding, and rolling motion. Their activities are freestyle competitions such as inline skating, street basketball, and snowboarding, which are often associated with a youthful subculture. Performance and rankings are no longer important. The esthetic, feel-good, and ambient effects are what matter.

Extreme sports have grown in popularity as a way for people to break free from the laws and regulations of others and form their own renegade groups. Even these games, though, are part of an evolutionary phase. Extreme sports tend to move from subculture to mainstream as they engage with sports institutions, media, equipment, and clothing manufacturers, and the tourism industry. Hoffer (2002) proposes that the cycle is perpetuated since, when the process of commodification intensifies, mechanisms and rules emerge to ensure that the operation is handled in a way that enables commodification before some community breaks apart to start something new.

Snowboarding is a clear example of this method. In response to the prevailing culture of alpine skiing, it arose as a subculture sport. Initially, it was distinguished by a non-traditional perspective on sport. Nonetheless, the initial progressive essence of snowboarding has been slowly moderated by commodification pressures. ESPN's decision to make snowboarding a commercial television product exemplifies this point. The inclusion of snowboarding in the 1998 Winter Olympics was another turning point in the transformation of snowboarding from a subculture to a mainstream sport.

8.5.6. The Evolutionary Dynamics of Sport Tourism

Tourism cycles also impact the development of sport. Golf, for example, has been introduced as a tourism development strategy in many hot climate destinations over the last two decades. Opportunities to participate in the sport of golf have been extended to and taken up by local residents as a result of this development. There is also a reciprocal relationship in that tourism allows for the popularization of leisure activities. They have evolved into formally organized sporting activities as their popularity has grown. Some even advanced from recreational activities to Olympic disciplines. Beach volleyball and snowboarding are two excellent examples.

Tourism not only introduces sport to new areas, but it also fosters sport innovation. Figure 8.5 depicts how change occurs in sport in both recreational and competitive settings, with the former being more conducive to great breakthrough than the latter. External trends in the economy, politics, society, technology, and the natural environment have an impact on both settings. External environments are frequently the source of sport innovations. Recreational sports are more conducive to innovation because most leisure and tourism settings encourage experimentation. Keller also contends that the change in location and uninterrupted free time that tourists enjoy while on vacation create an environment that is conducive to innovation in sports pursuits.

The emphasis in competitive environments is on success in terms of known physical skills and rules. These are built in such a way that fundamental change is impossible. Leisure trends that are peculiar to particular geographic regions form major innovations in recreational sport settings. In comparison to the lack of restrictions in a recreational setting, a number of sporting organizations serve as transition gatekeepers in competitive sport environments. Significant advances and gradual improvement in sports are disseminated spatially and institutionally through the broadcast of entertainment and news programs through a variety of media, as well as through the marketing campaigns of sporting goods manufacturers. Tourists are also important agents of creativity and dissemination. They bring new sporting preferences to tourist destinations, and as a result, they might be introduced to new sports while they are there.

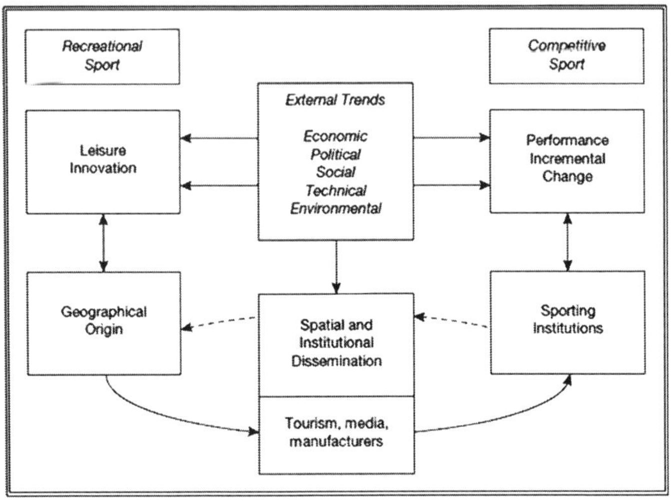

Figure 8.5. Innovation in sport.

Source: https://www.google.co.in/books/edition/Sport_Tourism_Development/FU1dIQ64yoEC?hl=en&gbpv=1&dq=Sport+Tourism&printsec=frontcover.

8.6. MAJOR TRENDS AFFECTING SPORT TOURISM

Strategic planning is an important aspect of sport tourism growth, albeit one fraught with difficulty. Trends in sport and tourism, both of which are ultimately discretionary practices, are among the most difficult to predict. Trend analysis, on the other hand, provides general insights into likely future scenarios in sport and tourism demand.

A number of assumptions have been made about sport tourism. Most experts predict that demand for sport and tourism experiences will continue to rise. This expansion, however, is unlikely to be a simple linear extension of current sport tourist participation trends. Table 8.3 outlines the latest developments in sport that are impacting sport tourism. According to Bourdeau et al. (2002), the fundamental trend in sport is diversification, with a turn toward individual sport. New sporting opportunities will continue to develop in urban environments, but peripheral areas will also gain prominence as sport danger zones.

Table 8.3. Key Trends in Sport Tourism

Trend	Description
1. The increasing development of individual sports as opposed to collective sports	This view is consistent with a societal change in mentality towards neo-individualism and personal development. Participants may gravitate towards other individuals who share their sporting passion, but this will likely be done outside of the traditional sporting institutions.
2. Diversification of sports participation models	People will generally be willing to try a broader range of sports activities than they have in the past. Similarity, sports disciplines will tend to open up their membership to a wider spectrum of participants.
3. Exaggerated segmentation of sports disciplines	An increasing variety of sport hybrids and specializations will emerge as new sporting experiences are sought.
4. Adaptation of sports activities to the constraints of urban life	The experiences associated with outdoor adventure sports will be simulated in the city where they are accessible to urbanities constrained by time from pursuing these activities in natural settings. A recent example is the emergence of indoor climbing facilities in urban areas.
5. Development of a mythology of adventure in a natural environment	Natural environments will grow in significance in real and symbolic terms as places where sport tourists confront uncertainty, risk and destiny.

These trends are embedded in wider tendencies in the larger sense in which sport and tourism occur. Economic, political, cultural, and technical developments shape the context for sport and tourism participation.

8.6.1. Economic Trends

Globalization is perhaps the most powerful economic force to emerge from the second half of the 20th century. It has placed further emphasis on the commodification of sports and tourism. The trend toward the convergence of travel, leisure, sport, and entertainment is particularly important. This is particularly true in elite organized sports, where a trend toward professionalism and show business is already visible.

The media has a significant impact on this operation. Sport has long been associated with popular culture, since the golden days of newspapers, radio, and television. However, it is the last of these types of media that is thought to have the most influence on elite sport. From the start, there was fear that televised coverage of sports would cause viewers to abandon live-action in favor of watching sports from the comfort of their own homes. The belief that sports revenues would change from conventional sporting institutions to broadcasters was at the heart of this concern.

Sport economics was based on the premise of persuading large numbers of people to leave their homes, travel to enclosed sporting venues, and pay for entry to watch professional athletes participate in different types of organized, physical competition.

Broadcasters have received considerable financial incentives in the past, but they have also provided major financial benefits for the owners and managers of televised sports, as well as the locations where these sports take place. Pro sports television revenues now greatly outnumber gate revenues. Despite the shifting economic background, there are still complaints that the media has subverted sport for its own ends, eroding the dignity of these sports in the process. Rowe, for example.

Television has gradually placed pressure on sports to be played at times suitable to broadcast schedules and to change rules in order to ensure outcomes, prevent events from running too long, and overcome any boring passages that could tempt viewers to reach for the dial (later the remote control). The global spread of sports broadcasting has imposed extreme pressures on sport, such as requiring that live sports be shown wherever possible at a time suitable for the largest and most lucrative TV markets.

Sport tourism faces significant challenges as a result of interactive technologies such as pay-per-view television, the Internet, and video games. Some writers have hypothesized that the increasingly immersive experience of watching sports from the comfort of one's own home may potentially necessitate paying spectators to attend televised games in order to create an enjoyable atmosphere in the sporting venue. The cumulative benefits of the on-site experience for sport visitors must clearly outweigh the trip costs. It is also essential for sport tourism managers to fight for the preservation of the characteristics that distinguish sport. Sport as entertainment is distinct from pure entertainment, and can be a very strong tourist attraction. Many tourist destinations benefit from the intrinsic authenticity of sport, which could be lost if the essence of the activity changes to staged entertainment.

The media's influence is not limited to professional sports. Subculture sports are also closely associated with the media. In today's postmodern culture, the professional press plays a critical role in introducing participants to strategies, equipment, cultural codes, and languages that form the foundations of the sports tribes' identities. This medium is thus of special interest to sport tourism managers because it determines where members of the subculture travel in search of their sporting passions.

8.6.2. Political Trends

Political dynamics are driven by shifting power and control balances. The rise of free-trade policies highlights shifting power dynamics around the world. The economic benefits of free trade are unlikely to be distributed equitably.

According to Hall, there will be increased tension between developing and developed countries about global economic development policies as it becomes clear to a significant proportion of the developing world's population that they will never be able to live Western lifestyles leading to population and resource constraints.

The consequences of such feelings and other political grievances are already being felt in the form of protest movements that have direct ramifications for tourism and sport. Terrorism is not a modern threat to tourism, nor is it foreign to sports. The fatal attack on Israeli athletes at the Munich Olympic Village in 1972 was perhaps the most significant terrorism incident in a sport tourism context. Nonetheless, considering the high visibility of sporting events as terrorist targets, there have been relatively few politically motivated attacks on sporting events. Indeed, it has been proposed that sport tourism was one of the most resilient forms of tourism in the wake of the 11 September 2001 terrorist attacks in America.

The most obvious assumption is that major sporting events would necessitate high standards of defense. There may also be a preference for off-site spectatorship over on-site consumption of the event. Active sports with dispersed spatial patterns of participation can gain popularity in comparison to major event sport tourism. Similarly, as conventional patterns of event sport tourism consumption change, nostalgic tourism may increase.

8.6.3. Socio-Demographic Trends

Socio-demographic changes are also expected to have a major impact on the future of sport tourism. The developed world's aging population is especially important. This pattern, for example, has had a direct effect on downhill skiing.

People born between 1946 and 1964 account for nearly a third of the North American population, and they began shifting to more gentle winter sports like snowshoeing and cross-country skiing as their aging bodies could no longer withstand the rigors of alpine runs. High-tech computer-designed skis and equipment, which make it safer and easier for even the most seasoned outdoor enthusiasts to learn or continue to enjoy the sport, have helped to slow the trend in recent years.

Table 8.4. A Comparison of Modern and Postmodern Sport

Dimension/Component	**Modern Sport**	**Postmodern Sport**
Game structure	Rules are sacred	Rule modification and experimentation
Team leadership	Conservative	Adventurous
Values and customs	Amateurism, respect for authority, character building	Professionalism, innovation
Organization and management	Central control	Diffusion of authority
Financial structure	Gate receipts	Sponsorship, television rights, gate receipts, sport as business
Venues and facilities	Basic seating at stadia	Customized seating, video support
Promotion	Limited	Extensive
Viewing	Live match attendance	TV audiences dominate
Spectator preference	Display of traditional craft	Eclectic blend of entertainment
Fan loyalties	Singular and parochial loyalty	Multiple loyalties-all spatial scales
The sports market	Undifferentiated mass market	Fragmented and niche markets
Coaching and training	Rigid, repetitive practices	Blend of science and naturalistic practices-variety

Faced with this trend, sport tourism operators must adapt their products to meet the needs of their markets. This change would entail a transition from physically demanding hard adventure activities to less physically demanding soft adventure outdoor sports activities. An aging population may become more health-conscious and, as a result, seek out sports activities that will help them maintain and possibly restore their health rather than jeopardize it.

The transition from a modern to a postmodern society is also influencing the sense of sport tourism. At one stage, the rejection of the welfare state and controlled economies in favor of competition, free trade, and globalization has economic roots. Today's culture is defined by niche markets, individualism, flexibility, time fragmentation, emerging technology, creative communication networks, and commercialization. The role of location in postmodern sport is also evolving. Local tribal loyalties dependent on the home team have largely given way to attachments to corporate identities or brands. Table 8.4 summarizes the evolution of professional sports from the early and postmodern periods. According to Murray and Dixon, the rise of instant sports is the product of the transition from modernity to postmodernity. This is consistent with Western society's focus on market orientations over citizenship orientations and unstructured sports over organized sports.

8.6.4. Technological Trends

Technological advancements have irreversibly altered the face of sport and tourism. They have enhanced sports performances as well as tourism experiences. Furthermore, technological advancements have blurred the distinction between sport and tourism. For example, the evolution of the Internet has resulted in the creation of a plethora of sport-related web sites ranging from the static provision of basic information about a sport to interactive sport sites that can serve as the foundation of a leisure experience in their own right. Virtual reality and cyberspace are having a direct effect on how people spend their leisure time. The degree to which sport experiences in cyberspace can substitute for sport experiences in real space is unknown. Live online sports commentaries, online sports gambling, and real-time audience polls during sport broadcasts, and instant progress reports, and live digital video photos of sports contests are recent examples of sport interactions that take place in cyberspace. On another level, many video games closely resemble concepts of sport, with the exception of the physical activity requirement, but this does not have to be the case. Physical activity should be integrated into these games on a far larger scale than it is now. Transportation technological advancements in the past have played an important role in tourism growth, and they are likely to continue to do so in the future. The expected arrival of a new generation of long-haul, wide-body, and multi-level jet aircraft would expand access and strengthen hub-and-spoke transportation trends. Continued advancements in space exploration would expand access to weightless worlds, potentially spawning a new wave of sports. Similar advances in marine ecosystems are likely to open up a slew of new opportunities for sport tourism in that field.

A number of environmentally controlled sports facilities, such as Toronto's Skydome with its retractable roof, have recently been installed. This development extends beyond major spectator sports to sporting events that were previously restricted to the outdoors and wilderness areas.

With free climbing, water slides, hydrospeeding, golf training, and skateboarding, sports-oriented theme parks are all the rage. There are ski domes in the United Kingdom and the Netherlands where beginners can test their skiing abilities and seasoned skiers can get in shape before the season begins. Leisure and theme parks, in general, do not depend on any [natural] landscape. They are, in reality, genuine industrial zones for sporting entertainment.

These facilities are typically built-in urban areas, where participants have convenient access in terms of both time (i.e., they can ski after work on a weekday) and space (i.e., they can participate in their preferred sporting activity close to home). If these urban-based artificial sporting environments were ideal replacements for the same sports in natural settings, it is reasonable to expect fewer visits to outlying areas. In fact, these simulated environments are not exact replicas. They will, in reality, act as demand shifters, as more people are exposed to the sport and may ultimately search out conventional activity sites. The initial natural settings are likely to remain the chosen venue for the vast majority of participants in the near future.

For example, the evolution of the Internet has resulted in the creation of a plethora of sport-related websites ranging from the static provision of basic information about a sport to interactive sports sites that can serve as the foundation of a leisure experience in their own right. Virtual reality and cyberspace are having a direct effect on how people spend their leisure time. The degree to which sports experiences in cyberspace can substitute for sports experiences in real space is unknown. Live online sports commentaries, online sports gambling, and real-time audience polls during sports broadcasts, and instant progress reports, and live digital video photos of sports contests are recent examples of sports interactions that take place in cyberspace. On another level, many video games closely resemble concepts of sport, except for the physical activity requirement, but this does not have to be the case. Physical activity should be integrated into these games on a far larger scale than it is now.

8.6.4.1. Management Implications and Opportunities

When attempting to affect the life cycle of a tourist destination, sport tourism managers should understand the cycles of various sporting events. It is critical to avoid sacrificing the nature of sporting activities, as doing so may result in the sport stagnating or declining, as well as a loss of competitive advantage at a destination. It should also be acknowledged that the media plays an important role in the life cycles of sporting events. Because of technological advancements and the growing sophistication of sports broadcasting, watching sports from the comfort of one's own home is becoming a more competitive alternative to attending live sports. On-site spectatorship can go beyond merely keeping up with the immersive media technologies available to at-home viewers. The intimate connection between athletes and fans, as well as the unique experience of location, must be preserved and enhanced. Sport managers must handle spectators as though they are an integral part of their sport, which they are, and provide opportunities to encourage spectators.

Given a large shift away from conventional sports and toward individualized activities such as fitness activities and extreme sports, active sport tourism appears to be a significant area of potential development. Sport tourism managers must keep a close eye on the development of new sports and gain a thorough understanding of the motivations that drive participation in these activities. The life cycles associated with them should be carefully observed, with special emphasis placed on whether the sport tends to have a long or short life cycle.

Managers of sport tourism should establish different plans for alternative and conventional sports. Extreme sports athletes, for example, can travel more independently than group travelers due to their freedom. Since the travel experiences pursued by extreme athletes vary from those sought by mainstream athletes, they are more likely to be enticed by different advertising ads, use different modes of transportation or fare systems, stay in different types of accommodations, and behave differently during their stay at a destination. Sport tourism executives who gain insights into the specific needs of these niche markets would be well placed for success. Nostalgia tourism is also a significant resource for sport tourism managers. Emotion is at the core of nostalgic memory. Visitors to sports halls of fame, high-profile sporting venues, and fantasy programs seek more than just facts. They want to have good emotional experiences. By researching contemporary interpretive approaches and drama, sport tourism managers will learn a lot about how to evoke this response. Creating economies of scale in terms of infrastructure and superstructure can be achieved by creating immersive sports halls of fame/museums at the actual stadia. More importantly, synergies between the hall of fame/museum and the atmosphere of the sporting venue can be created. These halls of fame/museums in new facilities help to bind the new sporting venue to its sporting heritage. Fantasy sports programs are only constrained by the imaginations of sport tourism executives. The chance to drink ambrosia with the gods or to fulfill one's sporting dreams is a powerful option. Programs that include these opportunities are likely to be successful in the mid-to-high-end sport tourism markets. Managers would have to strike a balance between imagination and reality that is appropriate to paying customers.

Bibliography

1. All Answers Ltd., (2018). *Impact of Major Sports Events on the Economy*. Retrieved from: https://ukdiss.com/examples/literature-review-on-economic-impact-studies.php?vref=1 (accessed on 12 June 2021).
2. Allen, J., et al., (2011). *Festival and Special Event Management* (5th edn.). John Wiley & Sons: Brisban.
3. Amis, J., & O'Brien, D., (2001). Organizational theory and the study of sport. In: Parkhouse, B. L., (ed.), *The Management of Sport: Its Foundations and Application*. New York: McGraw-Hill.
4. Aref, F., Gill, S., & Farshid, A., (2010). Tourism development in local communities: As a community development approach. *Journal of American Science, 6*(2), 155–161.
5. Baade, R. A., (1996). Professional sports as catalysts for metropolitan economic development. *Journal of Urban Affairs, 18*(1), 1–17.
6. Baim, D. V., (1994). *The Sports Stadium as a Municipal Investment*. Greenwood Press, Westport.
7. Bowdin, G., Allen, J., O'Toole, W., Harris, R., & McDonnell, I., (2006). *Events Management*. Elsevier Ltd, Great Britain.
8. Buerki, R., et al., (2003). Climate change impacts on the tourism industry in mountain areas. *1st International Conference on Climate Change and Tourism*. Djerba.
9. Burns, G., (2004). Anthropology and tourism: Past contributions and future theoretical challenges. *Anthropological Forum, 14*(1), 5–22.

10. Burns, J. P. A., Hatch, J. H., & Mules, F. J., (1986). *The Adelaide Grand Prix: The Impact of a Special Event, The Centre for South Australian Economic Studies*. Adelaide.
11. Coalter, F., & Allison, M., (1996). *Sport and Community Development*. Research Digest no. 42, Scottish Sports Council, Edinburgh.
12. Coalter, F., (1988). *Sport and Anti-Social Behavior: A Literature Review*. Research Report No. 2, Scottish Sports Council, Edinburgh.
13. Coalter, F., Allison, M., & Taylor, J., (2000). *The Role of Sport in Regenerating Deprived Urban Areas.* Centre for Leisure Research, University of Edinburgh, the Scottish Executive Central Research Unit, Edinburgh.
14. Crompton, J. L., (1995). Economic impact analysis of sports facilities and events: Eleven sources of misapplication. *Journal of Sport Management, 9*(1), 14–35.
15. Dabscheck, B., (2003). Paying for professionalism: Industrial relations in Australian rugby union. *Sports Management Review,* 105–125.
16. Deane, J., (1998). Community sports initiatives: An evaluation of UK policy attempts to involve the young unemployed-the 1980's action sport scheme. In: *Sport in the City: Conference Proceedings* (Vol. 2, pp. 140–159). Sheffield, Sheffield Hallam University, Sheffield.
17. Department of Culture, Media, and Sport, (1999). *Policy Action Team 10: Report to the Social Exclusion Unit-Arts and Sport.* HMSO, London.
18. Department of the Environment, Transport, and the Regions, (2000a). *Towards an Urban Renaissance the Report of the Urban Task Force Chaired by Lord Rogers of Riverside*. https://www.worldcat.org/title/towards-an-urban-renaissance-final-report-of-the-urban-task-force-chaired-by-lord-rogers-of-riverside/oclc/630109958 (accessed on 12 June 2021).
19. Gammon, S., & Kurtzman, J., (2002). *Sport Tourism: Principles and Practice*. Eastbourne: Leisure Studies Association (Segment 2).
20. Glyptis, S., (1989). *Leisure and Unemployment*. Open University Press, Milton Keynes.
21. Gratton, C., (1999). *Sports-Related Industry Study: Final Report*. Manchester City Council and Northwest Development Agency, Manchester.

22. Hall, C. M., (1992). *Hallmark Tourist Events: Impacts, Management, and Planning*. Belhaven Press, London.
23. Hall, C. M., (2001). Imaging, tourism, and sports event fever: The Sydney Olympics and the need for a social charter for mega-events. In: Gratton, C., & Henry, I. P., (eds.), *Sport in the City: The Role of Sport in Economic and Social Regeneration*. Routledge, London.
24. http://feaa.ucv.ro/AUCSSE/0036v2-024.pdf (accessed on 12 June 2021).
25. http://www.umdknes.com/knes287resources/Readings/10/R10.pdf (accessed on 12 June 2021).
26. https://core.ac.uk/download/pdf/234696865.pdf (accessed on 12 June 2021).
27. https://tourismnotes.com/rural-tourism/ (accessed on 12 June 2021).
28. https://www.dailyexcelsior.com/promoting-rural-tourism/ (accessed on 12 June 2021).
29. https://www.ierek.com/news/index.php/2018/07/25/urban-regeneration-world/ (accessed on 12 June 2021).
30. https://www.oecd-ilibrary.org/docserver/tour-2010-4-en.pdf?expires=1618813861&id=id&accname=guest&checksum=272914AB387046C713B41AA801C6403D (accessed on 12 June 2021).
31. https://www.oecd-ilibrary.org/docserver/tour-2010-4-en.pdf?expires=1619060064&id=id&accname=guest&checksum=ECFD4EFB1B24656AFAAA4305DC9B0BF4 (accessed on 12 June 2021).
32. https://www.researchgate.net/publication/236108688_Sustainable_Tourism_Development (accessed on 12 June 2021).
33. https://www.sportskeeda.com/sports/detailed-look-impact-sports-environment (accessed on 12 June 2021).
34. https://www.ukessays.com/essays/tourism/the-development-of-sport-tourism-industry-tourism-essay.php (accessed on 12 June 2021).
35. Kurtzman, J., (2000). *The PARRC Continuum-Sport Tourism Potentials Identified.* Occasional Papers, Sport Tourism International Council.
36. Kurtzman, J., (2001). Sports tourism futures: Presentation. *Leisure Studies Association Conference*. Luton: University of Luton.
37. Mathieson, A., & Wall, G., (1982). *Tourism: Economic, Physical, and Social Impacts*. New York: Longman Scientific and Technical.

38. McKercher, B., (1993). Some fundamental truths about tourism: Understanding tourism's social and environmental impacts. *Journal of Sustainable Tourism, 1*(1), 6–16.
39. Smith, M., (2009). *Issues in Cultural Tourism Studies* (2nd edn.) London: Routledge.
40. Tum, J., Norton, P., & Nevan, W. J., (2005). *Management of Event Operations,* 11. (Elsevier Ltd., Great Britain).
41. Turco, M., Riley, R., & Swart, K., (2002). *Sport Tourism (Chapter 10)*. Morgantown, WV: Fitness Technology, Inc.
42. Viljoen, J., & Dann, S., (2000). *Strategic Management*. Frenchs Forest, NSW: Longman.
43. Witt, S. F., (1991). Tourism in Cyprus, balancing the benefits and costs. *Travel Management, 12*, 37–46.

INDEX